ELECTRONIC FABRICATION

Gordon T. Shimizu

ELECTRONIC FABRICATION

Second Edition

Delmar Publishers Inc.®

NOTICE TO THE READER

Publisher does not warrant or guarantee any of the products described herein or perform any independent analysis in connection with any of the product information contained herein. Publisher does not assume, and expressly disclaims, any obligation to obtain and include information other than that provided to it by the manufacturer.

The reader is expressly warned to consider and adopt all safety precautions that might be indicated by the activities described herein and to avoid all potential hazards. By following the instructions contained herein, the reader willingly assumes all risks in connection with such instructions.

The publisher makes no representations or warranties of any kind, including but not limited to, the warranties of fitness for particular purpose or merchantability, nor are any such representations implied with respect to the material set forth herein, and the publisher takes no responsibility with respect to such material. The publisher shall not be liable for any special, consequential or exemplary damages resulting, in whole or in part, from the readers' use of, or reliance upon, this material.

Delmar Staff
Associate Editor: Cameron O. Anderson
Project Editor: Carol Micheli
Managing Editor: Gerry East
Production Coordinator: Larry Main
Design Coordinator: Susan Mathews

For information, address Delmar Publishers Inc.
2 Computer Drive West, Box 15-015
Albany, New York 12212

COPYRIGHT © 1990
BY DELMAR PUBLISHERS INC.

All rights reserved. Certain portions of this work copyright © 1986. No part of this work covered by the copyright hereon may be reproduced or used in any form or by any means—graphic, electronic, or mechanical, including photocopying, recording, taping, or information storage and retrieval systems—without written permission of the publisher.

Printed in the United States of America
Published simultaneously in Canada
by Nelson Canada
a Division of the Thomson Corporation

10 9 8 7 6 5 4 3 2

Library of Congress Cataloging-in-Publication Data

Shimizu, Gordon T.
   Electronic fabrication/Gordon T. Shimizu.—2nd ed.
     p.   cm.
   Includes index.
   ISBN 0-8273-3519-9.—ISBN 0-8273-3520-2 (instructor's guide)
   1. Electronic apparatus and appliance—Design and construction.
I. Title.
TK7870.S484   1990   89-7796
621.31'042—dc20   CIP

Contents

Preface   viii

SECTION 1: INTRODUCTION   1

Chapter 1: Safety   2
Laboratory Safety Procedures / 2
Electrical Shock / 2
Electrostatic Discharge / 3
Classification of Fire and Extinguishing Techniques / 8
General Safety Rules / 8

Chapter 2: Basic Concepts   11
Basic Electronic Hand Tools / 11
Soldering Basics / 13
Conductors and Cables / 17
Introduction to Printed-Circuit Boards and Soldering / 18
Schematic Diagrams and Component Identification / 19

SECTION 2: ELECTRONIC COMPONENT SYMBOLS   25

Chapter 3: Graphic Symbols of Common Electronic Components   26
Unit 1 Basic Passive Components / 26
       Resistors • Capacitors • Coils or Inductors • Transformers
Unit 2 Electronic Devices / 38
       Vacuum Tubes • Semiconductor Devices
Unit 3 Contacts, Contactors, Switches, and Relays / 51
       Mechanical Switches • Switch Contact Identification • Wafer Switch • Pushbutton Switches • Relays
Unit 4 Connection Devices / 58
       Connector Requirements • Connectors • Connector Symbols
Unit 5 Sources of Electricity / 63
       Direct Current (DC) Sources • Alternating Current (AC) Sources
Unit 6 Protection Devices / 64
       Fuses • Circuit Breakers
Unit 7 Lamps and Visual Signaling Devices / 66
       Lamps
Unit 8 Acoustical Devices / 66
       Speakers • Headsets and Earphones • Microphones and Other Sound Transducers
Unit 9 Rotating Machinery / 68
Unit 10 The Transmission Path / 69
        Wires and Cables • Cables
Unit 11 Ground Symbols / 70

Unit 12 Component Combinations / 70
   Coil and Transformer Combinations • Packaged Electronic Networks
Unit 13 Logic Symbols / 73
   The AND Function • The OR Function • The NOT and NAND Functions •
   The NOR Function • EXCLUSIVE OR Function • Other Logic Symbols
Unit 14 Miscellaneous Symbols / 76
   Antennas • Crystals

SECTION 3: ASSIGNMENTS 79

Assignment 1: Familiarization with Basic Electronic Hand Tools 80
Exercise 1: Wire Stripping / 80
Exercise 2: Common Electrical Wire Splices / 82

Assignment 2: Soldering Wire Terminations 85
Exercise 1: Tinning Stranded Wire / 86
Exercise 2: Soldering Wire Splice Samples / 87
Exercise 3: Developing Solder Feeding Skill / 88

Assignment 3: Connecting Wire and Component Leads to Terminal Strips 92
Exercise 1: Wire Crimping and Component Mounting / 92
Exercise 2: Soldering the Crimped Terminations / 97
Exercise 3: The Desoldering Process / 97

Assignment 4: Coaxial and Shielded-Pair Cable Assembly 100
Exercise 1: TSP (Twisted-Shielded-Pair) Cable Assembly / 100
Exercise 2: Coaxial Cable Assembly / 103
Exercise 3: Cable Inspection and Ohmmeter Testing / 106

Assignment 5: Printed-Circuit Board Assembly Techniques 108
Exercise 1: Component Identification and Mounting / 108
Exercise 2: Assembly of a Double-Sided PCB / 114
Exercise 3: Component Replacement and Repair of the Assembled PCB / 115

Assignment 6: Schematic Reading 118
Exercise 1: Identifying Graphic Symbols / 118
Exercise 2: Developing Schematic Diagrams / 122

SECTION 4: PROTOTYPE CONSTRUCTION PROJECTS 129

Project 1: Chassis Fabrication 130
Exercise 1: Interpreting the Sheet Metal Layout Drawing / 130
Exercise 2: Sheet Metal Layout Technique / 134
Exercise 3: Operating Sheet Metal Equipment to Complete Chassis Fabrication / 136

Project 2: Chassis Assembly — 140
Exercise 1: Identification of Hardware and Components / 140
Exercise 2: Chassis Assembly Procedure / 144

Project 3: Chassis Wiring Procedure — 153
Exercise 1: Identification of Wiring Diagrams / 153
Exercise 2: Schematic Reading and the Wiring Procedure / 158

Project 4: Continuity/Voltage Tester — 168
Exercise 1: Component Identification / 168
Exercise 2: Preparation of the Control Panel / 170
Exercise 3: PCB Design and Fabrication / 175
Exercise 4: Final Assembly and Testing / 182

Project 5: Blinking LED Circuit — 188
Exercise 1: Circuit Breadboarding Technique / 189
Exercise 2: PCB Planning Process / 191
Exercise 3: Artwork Development / 194
Exercise 4: PCB Fabrication Method / 195
Exercise 5: PCB Assembly and Testing / 200

Project 6: Adjustable, Bipolar, Regulated Power Supply — 202
Exercise 1: Circuit Introduction and Component Identification / 203
Exercise 2: Chassis Planning and Fabrication / 206
Exercise 3: PCB Planning and Fabrication / 209
Exercise 4: PCB and Final Chassis Assembly / 217
Exercise 5: Testing Procedure and Spec Sheet Development / 219

Appendixes — 229
Appendix A. Symbol Reference Guide / 229
Appendix B. Wire Wrapping: The Technology, Methods, and Standards / 242
Appendix C. Standard Resistor Color Code / 247
Appendix D. Industrial Rectifier Outlines / 251
Appendix E. Transistor Outlines / 252
Appendix F. SCR Outlines / 259
Appendix G. Triac Outlines / 261
Appendix H. Discrete LED Indicator Outlines / 262
Appendix I. Integrated Circuit Standard Package / 263
Appendix J. Electrical Safety Exam / 264

Index — 268

Preface

Electronic Fabrication, second edition, is a one-semester fabrication text that encompasses all the necessary skills and techniques that are needed by the electronics technician. These skills include sheet metal layout techniques, chassis fabrication, schematic interpretation, printed circuit board design, soldering techniques, final assembly and testing procedures.

The objectives of the first edition of this text were to combine the content of a two-semester electronic fabrication course into one book, covering the basic skills and then applying them to practical and interesting projects. These objectives were a success and, hence, a second edition of the text became a necessity.

The second edition has been expanded to include a more comprehensive safety chapter which explores electrostatic discharge (ESD). Better coverage of hand tools and schematic reading have also been included. A new project on chassis fabrication allows the student to learn the basic skills and then proceed through every stage of fabrication that will be needed in the workplace. Six comprehensive and sequential projects ensure that the student will experience almost every situation that is likely to arise in the work environment.

ACKNOWLEDGMENTS

I am grateful to all my colleagues on the electronics staff for their support and input in this project. I am also grateful to the following people and colleges who reviewed the manuscript and provided me with pertinent suggestions for its improvement.

Alexander Padgett
Karl Miller
Roger Eddy
Sinclair Community College
Thief River Falls Technical Institute
North Central Technical Institute
College of Aeronautics
Del Mar College
Texas Southmost College
Eastern Arizona College

SECTION 1
INTRODUCTION

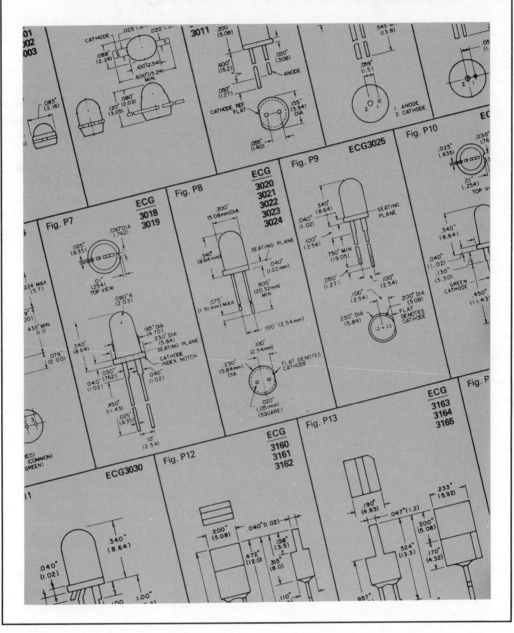

Chapter 1: Safety

LABORATORY SAFETY PROCEDURES

People working in industry know the importance of a safe working environment and safe working habits. Safety training programs are occasionally sponsored by unions, management, public agencies, and insurance companies. Despite these good efforts, accidents annually cause lost job time, painful injuries, and needless deaths.

Good safety habits are learned daily. As you begin your work in electronics, resolve now to learn and practice safe working habits in the laboratory.

Safe Attitudes

Laboratory working areas are for adults. Tricks, games, and general horseplay are not appropriate.

Safe Environment

Work areas must have proper power, ventilation, and light. Aisles should be open and clear. Storage areas should be kept clean and secure. The use of temporary extension cords, fans, heaters, gas, or water connections is discouraged. Maintain a neat and orderly work area.

First Aid Procedures

Even with good safety practices, someone may be injured. Your instructor and/or school nurse are trained in first aid procedures, but there are several general rules that you should follow:

1. Don't panic! Determine if there is any immediate danger to the injured person. Never move an unconscious person without cause. Lay the person flat. Keep the person warm to prevent shock. Never try to force liquids on an unconscious person. If the person is breathing normally, keep him or her still and comfortable until medical aid arrives.
2. Severe electrical shock or other types of accidents may interrupt breathing. A procedure such as artificial respiration or CPR (Cardiopulmonary Resuscitation) can be used to stimulate the breathing process. Check for a swallowed tongue before applying artificial respiration. This procedure should be administered by a trained person if possible, and continued until medical help arrives.
3. Report all injuries to the instructor immediately. Even minor cuts can become infected, and the best first aid supplies, nurses, and doctors cannot help an unreported injury.

ELECTRICAL SHOCK

One of the major hazards in the electronics field is electrical shock. Shock is caused by the passing of electric current through the body. Current flow is related to the voltage applied; therefore, the higher the voltage, the more serious the shock. Keep in mind, though, that shock produced by low voltage can be just as serious as shock produced by high voltage.

Following are the probable effects of different levels of current on the body:

.001 amp (1 mA)	Produces a shock that can be felt (mild tingling sensation).
.01 amp (10 mA)	Produces a severe, painful shock and can cause a loss of muscular control (can't let go phenomenon).
.1 amp (100 mA)	Produces a shock that can cause death if current lasts for a second or more.

Other possible effects of electrical shock are muscular paralysis, burns, cessation of breathing, unconsciousness, ventricular fibrillation, and cardiac arrest.

It is easy to see that the body is sensitive to relatively small values of current. In comparison, a common 100-

watt light bulb draws approximately 0.85 amp (850 mA) of current when connected to a 120-volt source—much higher than the 0.1 amp (100 mA) of current that can cause death.

All of these conditions, however, do not occur with every exposure to electrical shock. As stated before, factors vary. The following factors influence the effect of electrical shock on the body:

1. The intensity of the current
2. The frequency of the current
3. The current's path through the body
4. How long the current passes through the body
5. The element of surprise

Keep in mind that the amount of current flow through the lethal path, not necessarily the amount of voltage contacted, is the determining factor in the severity of a shock. The larger the current, the more dangerous and fatal is the shock.

ELECTROSTATIC DISCHARGE (ESD)

A comprehensive ESD control program should be implemented in all facets of the manufacturing chain due to the modern technological trend toward electronic devices that consume less power, require lower applied voltage, and are more susceptible to static electricity. Each contributing member of an electronic manufacturing plant should be educated in regard to this important topic that plagues today's modern electronic industries. An ESD control program should begin by providing proper administrative support. This administrative support individual or group should organize and set program requirements, provide the awareness and training program for all personnel in the manufacturing chain, and develop a monitoring system to insure that all safety guidelines are enforced.

The effects of ESD on electronic devices are not generally recognized because they are often masked by reasons such as electrical overstress due to transients other than static and failure categorized as a random, unknown, manufacturing defect, or another analysis label. Also, few failure analysis laboratories are equipped with the proper equipment and technology to detect failures due to ESD. Lack of implementation of ESD controls has resulted in high repair costs and excessive equipment downtime.

Static electricity is actually electrical charge at rest. The electrical charge is due to a transfer of electrons within a body known as polarization. Conductive charging occurs when electrical charge is transferred directly from one body to another. Some substances give up electrons readily; others tend to accumulate excess electrons. A body with an excess of electrons is charged negatively; a body with a deficit of electrons is charged positively. A surface may be charged by three means: triboelectricity, induction, and the principle of capacitance.

The most common means to charge neutral bodies is *triboelectricity*. Triboelectricity is an electrical charge generated by friction. When two solid substances are rubbed together, one gains electrons and the other loses electrons. If the substances are gases or liquids, a similar occurrence results when the flow is relative to one another. The magnitude of the charge depends on the size, shape, composition, and electrical properties of the substances. A charge in excess of 5 kV (5000 volts) may develop. A person can develop a significant charge simply by walking across a carpeted floor or removing a garment. The charged individual is a mobile, high-voltage source ready to discharge its supply of electricity to an uncharged or neutral body.

Electronic components are also able to acquire charges during processing. Consider, for example, integrated circuits packaged and shipped in plastic tubes. Charges can be developed on these devices from their movement inside the tube. When a charged device is removed from the tube, a rapid discharge can cause the device to fail.

A second and more subtle means of charging is by *induction*. Charging by induction occurs when an electrical conductor develops a charge as a result of its close proximity to a charged body. A good example that illustrates this charging principle occurs when a person handles a printed-circuit board or component wrapped in plastic, bubble-wrap packing material. The person handling the plastic material induces a charge onto the plastic which, in turn, transfers or induces a charge to the contents inside the wrap. When another person handles the wrap to remove the contents, a sudden discharge will result, causing ESD damage.

The third means of charging a neutral body employs the principle of *capacitance*. The familiar equation for charge, $Q = CV$ (charge equals capacitance times voltage), when solved for voltage is $V = Q/C$. An analysis of this equation shows that if the charge is constant and the capacitance decreases, the voltage will increase. Since capacitance is inversely related to distance between two conductive surfaces, a harmless low voltage in a circuit with a relatively low capacitance to ground can become a harmful voltage as the object is moved further from the ground plane. For example, when a circuit assembly on the floor or table is picked up, its relative capacitance decreases and voltage increases. When the assembly is grounded again, it will probably be found that it has been damaged by the large voltage generated when it was lifted from its original grounded position.

Typical prime charge sources commonly found in a manufacturing facility are listed in table 1-1. These sources are predominantly insulators and are typically synthetic materials. The voltages generated with these insulators can be extremely high because they are not readily distributed over the entire surface of the substance or conducted to another contacting substance. The conductivity of these insulators is increased by the absorption of moisture under high humidity conditions onto the otherwise insulating surface, creating a slightly conductive sweat layer that tends to dissipate static charges over the material surface.

TABLE 1-2 Typical Prime Charge Sources

OBJECT OR PROCESS	MATERIAL OR ACTIVITY
Work surfaces	• Waxed, painted, or varnished surfaces
	• Common vinyl or plastic
Floors	• Sealed concrete
	• Waxed, finished wood
	• Common vinyl tile or sheeting
Clothes	• Common clean room smocks
	• Common synthetic personnel garments
	• Nonconductive shoes
	• Virgin cotton
Chairs	• Finished wood
	• Vinyl
	• Fiberglass
Packaging and handling	• Common plastic (bags, wraps, envelopes)
	• Common bubble-wrap packing or foam
	• Common plastic trays, plastic tote boxes, vials, parts bins
Assembly, cleaning, test, and repair areas	• Spray cleaners
	• Common plastic solder suckers
	• Solder irons with ungrounded tips
	• Solvent brushes (synthetic brushes)
	• Cleaning or drying by fluid or evaporation
	• Temperature chambers
	• Cryogenic sprays
	• Heat guns and blowers
	• Sandblasters
	• Electrostatic copiers

Humans are prime sources of ESD for damaging components. Electrostatic charges generated by rubbing or separating materials are readily transmitted to a person's conductive sweat layer causing that person to be charged. When a charged person handles or comes in close proximity to an ESDS (Electrostatic-Static Discharge Sensitive) component, he can damage that part from direct discharge by touching or subjecting it to an electrostatic field. For test purposes, the ESD human model can be simulated by a test circuit as illustrated in figure 1-1. This test circuit, widely used in industry to represent a person for ESD testing, is applicable to all testing. Normally, component failure is defined as the inability of a part to meet the electrical parameter limits of its given specification. Any measurable change in a component's electrical parameter, due to an ESD, could indicate part damage and susceptibility to further degradation and subsequent failure.

Human capacitance can be as high as several thousand picofarads (pF) but is typically in the range of 50 pF to 250 pF. This variation in human capacitance is due to factors such as variations in the amount and style of clothing worn by personnel and also by the differences in the size and type of floor material of the work area. Human resistances can range from 100 ohms to 5 kilohms, but are typically between 1 kilohm to 5 kilohms for actions considered pertinent to holding or touching parts or containers of parts. Variations in human resistances are due to factors such as the amount of moisture, salt, and oils at the skin surface; skin contact area; and pressure. Table 1-2 shows typical electrostatic voltage levels generated by personnel in a manufacturing facility.

For voltage-sensitive ESDS components, a variation in the capacitance value in the test circuit will have little effect on its sensitivity. A decrease in human model resistance increases the voltage and power delivered to the part and causes the voltage level at which damage occurs to decrease. Therefore,

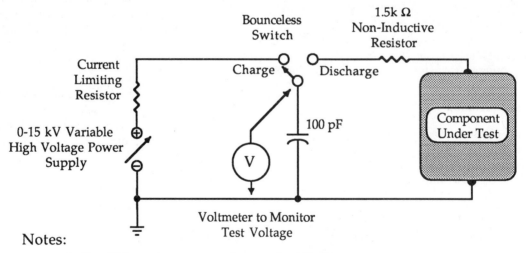

Notes:

1. Test Voltages are measured across the 100 pF capacitor. Test voltages should be within a tolerance of ± 5 %.

2. The capacitor is discharged throught the series 1.5kΩ into the component under test by switching to the discharge position on the bounceless switch for a time no shorter than required to decay the capacitor voltage to less than 1 % of the test voltage or 5-seconds, whichever is less.

FIGURE 1-1 ESD test circuit

TABLE 1-2 Typical ESD Voltage Levels

MEANS TO GENERATE STATIC CHARGE	VOLTAGE LEVELS (kV) AT % OF RELATIVE HUMIDITY	
	10-20%	65-90%
Walking across carpeted floor	35	1.5
Walking across vinyl floor	12	0.25
Employee at workstation	6	0.1
Vinyl envelope containing instrument manual	7	0.6
Picking up common plastic bag from workstation	20	1.2
Work chair padded with polyurethane foam	18	1.5

voltage-sensitive components can have a larger decrease in voltage damage level. The test circuit employing the 100 picofarads to 1.5 kilohms human model is considered to be a reasonable test circuit for determining the sensitivity of components to ESD.

Although equipment or instrument failures can occur during operation, catastrophic ESD failures can occur at any time. They can be the result of electrical overstress of electronic parts caused by ESD such as a discharge from a person or object, an electrostatic field, or a high-voltage spark discharge. Some catastrophic failures might not occur until exposure to an ESD has taken place, such as the case of marginally damaged parts that require normal operating stresses and operating time to cause the failure to become apparent.

ESD-related failure mechanisms include the following:

1. Thermal secondary breakdown
2. Metallization melt
3. Bulk breakdown
4. Gaseous arc discharge
5. Surface breakdown
6. Dielectric breakdown

The first three failures are power related; the last three are voltage dependent. Unencapsulated ICs and LSI MOS ICs have exhibited temporary failure due to gaseous arc discharge. This is a result of positive charges being deposited on the chip as a byproduct of gaseous arc discharge within the package between the lid and the substrate. The common failure mechanisms of film resistors are metallization melt and gaseous arc discharge. Bulk breakdowns are more evident in the piezoelectric crystal. These failure mechanisms apply to all semiconductor and microelectronic devices.

Thermal secondary breakdown, commonly called *avalanche degradation,* is caused by insufficient diffusion of heat from the power dissipation surfaces that form large temperature gradients within the material itself. This is due to the long thermal time constants of these materials compared to the short duration of the ESD pulse. These localized temperature gradients or hot spots can approach material melt temperatures that cause subsequent junction shorts due to melting.

Metallization melt failure occurs when EST transients increase the temperature of the component sufficiently to melt metal or the fuse bond wires. This type of failure usually occurs where the metal strips have a reduced cross section as they intersect oxide layers. As a result, a shunt current path is provided by the junction. This failure, however, requires larger power levels at higher frequencies compared to lower frequencies. Below 200 megahertz to 500 megahertz, the junction capacitance still presents a high impedance to currents, thereby shunting them around the junction.

Bulk breakdown is usually preceded by thermal secondary breakdown. This particular failure mechanism results from changes in junction parameters due to high concentrated temperatures within the area of the junction. Such high temperatures cause a change in junction parameters from a process known as *metallization alloying* or *impurity diffusion*. This usually results in the formation of a resistance path across the junction.

Gaseous arc discharge usually causes degradation of performance for components that contain closely spaced unpassivated thin electrodes. The arc discharge causes vaporization and metal movement, which is generally away from the space between the electrodes. In melting and fusing, the metal pulls together and flows or opens along the electrode lines. It does not move the thin metal into the inter electrode regions. Fine metal globules are evident in the gap regions but not in a sufficient amount to cause bridging. Shorting is not considered a major problem with unpassivated thin metal electrodes.

Surface breakdown results in a narrowing of the junction space charge layer at the surface due to a process known as *localized avalanche multiplication*. The destructiveness of this failure mechanism results in a high leakage path around the junction, thus nullifying the junction action. This effect, as well as most voltage-sensitive effects, depends on the rise time of the ESD pulse and usually occurs when the voltage threshold for surface breakdown is exceeded before thermal failure can occur. Another result of this failure is the occurrence of an arc around the insulating material, which is similar to metallization-to-metallization gaseous discharge. However, in this case, the discharge is between metallization and semiconductor.

Dielectric breakdown is due to voltage rather than power. When a potential is applied across a dielectric region in excess of the breakdown specification, a puncture of the dielectric material occurs. The breakdown potential (voltage) of an insulated layer is a function of the pulse rise time since time is required for avalanche to occur between the insulating material. This failure mechanism could result in either total or limited degradation of the component. For example, a part may "heal" from a voltage puncture if the energy of the pulse is insufficient to cause fusing of the electrode material at the punctured area. It will, however, usually exhibit lower breakdown characteristics or increased leakage current after such an event, but not immediate catastrophic component failure.

Different components are susceptible to ESD in various degrees. These variations are mainly due to different material

TABLE 1-3 Components Susceptible to ESD

PART CONSTITUENT	PART TYPE	FAILURE MECHANISM	FAILURE INDICATOR

MOS devices	• Discrete MOSFET
	• MOS IC
	• Semiconductors with metallization crossovers:
	   Digital ICs (bipolar and MOS)
	   Linear ICs (bipolar and MOS)
	• MOS capacitors
	   Hybrids
	   Linear ICs
		• Dielectric breakdown from excessive voltage and subsequent high current	Short (high leakage)

Semiconductor junctions	• Diodes (PN, PIN, Schottky) • Bipolar transistors • JFET (junction field-effect transistors) • SCR & Triacs • Digital and linear bipolar ICs	• Microdiffusion from microplasma secondary breakdown from excess energy or heat • Electronmigration (current filament growth) by silicon and aluminium diffusion	
Film resistors	• Thick film resistors • Monolithic IC-thin film resistors • Encapsulated film resistors	• Dielectric breakdown; voltage-dependent creation of new current paths • Joule heating energy-dependent destruction of minute current paths	Resistance change
Metallization strips	• Hybrid ICs • Monolithic ICs • Multiple finger overlay transistors	• Joule heating energy-dependent metallization burnout	Open strips
Field effect structures and nonconductive lids	• LSI and memory ICs employing nonconductive quartz or ceramic package lids, especially ultraviolet EPROMS	• Surface inversion or gate threshold voltage shifts from ions deposited on surface from ESD	Degradation of operation
Piezoelectric crystals	• Crystal oscillators • Surface acoustical wave devices	• Crystal fracture due to mechanical forces when excessive voltage is applied.	Degradation of operation
Closely spaced electrodes	• Surface acoustical • Thin metal unpassivated, unprotected semiconductors and microcircuits	• ARC discharge melting and fusing of electrode metal	Degradation of operation

composition of these parts and their architectural configuration. Table 1-3 lists several common component constituents that are sensitive to ESD failure. The table also summarizes the common failure mechanism of each part constituents.

The key to a properly created ESD control program includes electrostatic monitoring instrumentation. Without proper instrumentation, damaging charges cannot be detected. Electrostatic detectors include electrometer amplifiers, electrostatic voltmeters, electrostatic field meters, and leaf deflection electroscopes. Hand-held static meters can be used to detect both the potential and polarity of charges on any surface, including plastic films, tote boxes, clothing, and personnel. A limitation of most electrostatic meters is their response time. Most meters are incapable of responding to pulses with fast rise and short pulse widths. For measuring pulses with very fast rise and decay times, a high-speed storage oscilloscope can be used instead. Static level alarm systems are also available for constantly monitoring the levels of static electricity generated in a protected area. Some systems have multiple remote sensors that can monitor several stations simultaneously. Some systems also contain strip chart recorders that can provide a permanent record of the static levels within an area. In summary, when selecting any electrostatic detecting instrument, consider the following characteristics:

1. Sensitivity of instrument
2. Response time
3. Voltage range that can be measured
4. Accuracy range
5. Radioactive or electrical operation
6. Portability
7. Ruggedness
8. Simplicity of operation and readability
9. Accessory options (remote probes and strip chart recorder output)

The traditional attitude toward static electricity has been to deal with the problem at the specific location. Protection against the generation of electrostatic charges is the best method of ESD control. If materials do not generate electrostatic charges, no other action is required. One of the prime characteristics of materials in reducing the generation of static is *lubricity*. Lubricity is a measure of surface smoothness and the lubricating action of its surface moisture. Because triboelectric generation is a friction process, the higher the lubricity of the surface being rubbed, the lower the friction; hence the lower the charges being generated. Moisture on the surface of materials being separated provides progressive neutralization of opposite charges by furnishing a conductive path between the surfaces until separation is complete. Once a charge is generated, the distribution of that charge is dependent upon the resistivity and surface area of the material. The more conductive the material, the faster the charge is distributed. The greater the surface areas over which a charge is spread, the lower the charge density and the level of the residual voltage.

Conductivity is also a prime characteristic for providing protection against stationary or approaching charged bodies or people by limiting accumulation of residual charges. A polarization occurs gradually as the charged body or person approaches, limiting the voltage levels induced across the conductive surface. In conductive materials, these electrons move quite rapidly, resulting in low voltages applied across the ESD protective material. As the resistivity of the material increases, such as in the case of static dissipative and antistatic materials, the electrons move more slowly and higher voltages result. If the voltage of the charged body is high enough and approaches a tote box or tabletop, which is highly conductive, a spark will occur. The more conductive the material, the higher the probability of creating a spark and the higher the discharge current. Higher discharge currents conducted through a component increase its probability of failure. The tote box or tabletop should be conductive enough so that significant voltages will not be induced across the tote box or tabletop, but not be so conductive that a large spark discharge will occur.

Complete shielding from electrostatic fields or ESD high-voltage spark requires enclosing the item in a conductive material. Normally, the greater the conductivity of the enclosure, the greater the attenuation of the electrostatic field and ESD high-voltage spark induced within the enclosure.

A self-control program of ESD prevention requires the awareness and practice of some basic rules.

RULE 1 Treat all electronic components and accessories as static sensitive.

- Do not touch the leads or pins of components or circuit traces.
- Keep parts in original containers until needed.
- Before handling devices, discharge yourself by touching a grounded metallic surface such as a rack or cabinet. If available, use a wrist strap grounded through a 1-megohm resistor.
- Avoid sliding static-sensitive devices over any surface.

RULE 2 Handle all sensitive components and assemblies at static work stations.

RULE 3 Package components and assemblies properly for storage or shipping.

- Storage or shipping envelopes or containers should have approved, official warning labels.
- Use antistatic packing material when packaging parts for storage or transportation. Pack tightly to prevent any motion that could generate static electricity.
- Use antistatic tubes to store or ship ICs.
- Be sure that any equipment on wheels, casters, frames, or shelves is made of conductive material.
- Apply approved topical antistatic agents to surfaces that are potential sources of static generation by triboelectric effects.

ESDS components or assemblies that come in contact with workbenches and personnel should have EDS protective work surfaces over the entire areas where these items would be placed. Personnel ground straps are a necessary supplement to ESD protective workbench surfaces to prevent personnel discharging an ESD through an ESDS item to the workbench surface. Workbench surfaces should be connected to ground through a ground cable. The resistance in the benchtop ground cable should be located at or near the point of contact with the workbench top and should be high enough to limit any leakage current to 5 milliamperes or less. This consideration is dependent on the highest EDS voltage level within reach of grounded people and all parallel paths to ground such as grounded wrist straps, tabletops, and conductive floors.

A static-safe work station such as the one shown in figure 1-2 contains the following items and precautions:

1. A conductive table mat grounded through a 1-megohm resistor. Each mat should have two swivel connectors: one for the worker and one for the supervisors, inspectors, and so on.
2. A conductive floor mat that is grounded. Conductive heel straps should be used where walking is necessary and wrist straps cannot be worn. A new heel strap must be used each day. Shoes with conductive soles designed to be worn in antistatic environments can be worn as an alternate to heel straps.
3. Conductive wrist straps in contact with bare skin and connected to the swivel connector on the mat.
4. All metal equipment, such as soldering irons, workbenches, machinery, electrical equipment, and fixtures, must be grounded.
5. Each work station should contain one common ground. For example, the table mat and equipment must be connected to the same grounding point. A screw-on metal junction box or a properly grounded ac power line is a good place to attach a ground cable.
6. The work station must be clear of nonconductors such as plastic products, cardboard, work envelopes, candy wrappers, and synthetic mats.
7. Clothing must not touch components or assemblies. Short sleeves are preferred; if long sleeves are worn, they must be rolled up enough to prevent contact with

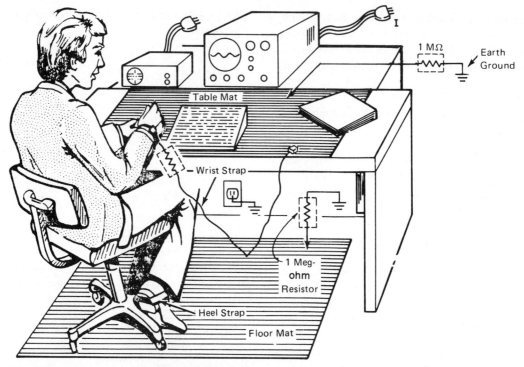

FIGURE 1-2 Static-free work station for board repair. All test equipment must be grounded.

or close proximity to sensitive parts. A long-sleeved smock or sleevelets made of ESD-protective material should be worn, if available.
8. Use approved containers, such as static-protective bags, conductive or antistatic trays, and antistatic tubes for ICs. Do not place paper or cards inside the containers.
9. Gloves should be made of cotton or antistatic material, not synthetic material.
10. Carts used to transport sensitive items should be covered by conductive mats and should have at least two conductive wheels.

CLASSIFICATION OF FIRE AND EXTINGUISHING TECHNIQUES

There are three categories of fire, each of which requires special extinguishing techniques. Regardless of the type of fire, if an extinguishing device is used, the nozzle should be pointed at the *source* of the fire, *not* at the *top* of the flame. Following are descriptions of extinguishing techniques.

CLASS A — Fires involving such combustible materials as wood, paper, or cloth. To extinguish this type of fire, *cool* it with a pump-type extinguisher containing water or soda acid. CO_2 (carbon dioxide) extinguishers may also be used.

CLASS B — Fires involving such flammable liquids as gasoline, kerosene, greases, thinners, finishes, or other solvents. To extinguish, *smother* the burning fuel. Foam and CO_2 extinguishers may be used.

CLASS C — Fires involving electrical equipment. To extinguish, use non-conducting extinguishers such as CO_2 or dry-powder extinguishers. If possible, disconnect the source of electrical energy.

CAUTION CO_2 *extinguishers do not always remove the heat from fires; therefore, flashback may occur.*

GENERAL SAFETY RULES

Lab Conduct and Safety Practices

1. Obey all posted warning signs. They are posted for your protection.
2. It is your responsibility to caution any individual who is violating any safety rules.
3. Work at a pace consistent with safety. Foolish hurry, such as rushing to complete a procedure, is dangerous.
4. If a piece of equipment or a tool is not working properly, inform your instructor or supervisor immediately.
5. Report any dangerous situation encountered in the lab to your instructor or supervisor immediately.
6. Any form of horseplay, including throwing any object, is dangerous and is forbidden at all times.

7. Consider every electrical circuit live until proven otherwise.
8. Be sure that your hands are completely dry before operating or working around electrical switches, plugs, or receptacles.
9. Never allow anyone to turn power on or off for you while working with machinery, or on any electrical circuits.
10. Learn the location of emergency switches, fire extinguishers, and emergency exits.
11. When lifting heavy objects, keep your arms and back as straight as possible, bend your knees, and lift with the more powerful leg muscles. Never lift heavy objects alone. Get some help.

Personal Protection

1. Eye protection must be worn at all times in areas where hazardous conditions exist.
2. Face shields or goggles should be worn when extra protection is required, such as when grinding, or working with caustic chemical substances.
3. When compressed air is used for cleaning, eye protection must be used. Direct chips, shavings, or dust away from other workers. Never allow the stream of air to come in contact with your body.
4. Fasten or remove loose clothing before operating any machinery. Roll long sleeves above the elbows.
5. Wear closed shoes when working in the lab. Open sandals can not be worn in the lab.
6. Wearing gloves is discouraged when working with power-driven machinery in the lab.
7. Long, loose locks of hair can easily be caught in revolving machinery, causing a severe scalp laceration. Long hair should be fastened to avoid any possibility of entanglement around the revolving shafts of a machine.
8. Wear protective clothing such as an apron and rubber gloves when working with chemicals.
9. Always wash hands with soap and water after working with materials that could cause any irritation to your skin.
10. Remove jewelry such as loose bracelets, rings, necklaces, chains, or other ornamental objects that could be a hazard around rotating machinery or exposed electrical circuits.

Housekeeping

1. Keep your immediate work area orderly and clean during working hours. Good housekeeping practices are an important aspect of safety.
2. Keep floors, aisles, and passageways clear of materials, electrical cords, and equipment.
3. Never leave tools where they may cause injury. Put them in the tool box, tray, case, or wall panel.
4. Store large stock material, such as sheet aluminum and chemicals, where it will not become an obstacle to people in the work area.
5. Eating and drinking are prohibited in the laboratory area at all times.
6. Route extension cords away from vehicular or foot traffic to avoid having them run over or stepped on. If vehicular or foot traffic is unavoidable, be sure that the extension cords are well protected.
7. Use a brush to clean benches and machines. The scraps may contain sharp or jagged particles that could seriously injure your hands.
8. Always keep doors to bench cabinets, drawers, and locker doors closed.
9. Keep tools and materials from projecting over the edge of benches or tables whenever possible.

Hand Tools

1. Keep your hands as free as possible of dirt, grease, or oil when using any tools or equipment.
2. Selection of the proper tool for a specific job is important. Use the right type and size hand tool, or proper equipment for the required procedure.
3. When carrying tools in your hands, keep the cutting edge or point directed towards the floor.
4. Small work pieces should be clamped to an appropriate holder or vise when using a hacksaw or screwdriver, or when soldering circuit boards.
5. Never use chisels, punches, or hammers with mushroomed heads. Mushroomed edges may become flying projectiles and seriously injure you or another individual.
6. Never use a file without a handle. Be sure that the handle is properly secured to the file.
7. When passing any sharp tool to another individual, position the tool so that the handle is received by the recipient.
8. Do not use tools with plastic handles near an open flame or a hot soldering iron.
9. Keep metal rules clear of live electrical circuits. When in doubt, use a plastic or wooden rule.
10. Disconnect all portable electrical tools and equipment from electrical outlets when changing cutting tools or drill bits, and when not using them.
11. When disconnecting any electrical tool, equipment, or appliance from an outlet, remove the attachment plug from the receptacle by pulling on the plug handle, not the cord.
12. All tools and equipment are rated according to their maximum performance capacity. There is a proper and safe way to use each tool and piece of equipment. If you are unsure of the proper use or

operation of a tool or piece of equipment, refer to the operating manual.

13. Never use a pair of pliers to cut two or more wires at the same time; the tool may be damaged or a person may be injured, particularly if power is applied.

Machine and Power Tools

1. Only the operator may start or stop the machine. After turning the switch off, the operator should wait until the revolving mechanism has come to a complete stop before leaving the machine.
2. All adjustments must be securely fastened before power is turned on.
3. Remove all cleaning implements such as rags or brushes, and any adjustment tools such as wrenches, screwdrivers, or other loose tools from the work area before power is applied to the machine.
4. Keep machinery and safety guards in their proper positions at all times.
5. Overloading or forcing a cutting or drilling tool into a workpiece is dangerous. Use only approved material furnished or approved by your instructor.
6. Be sure that you are the only individual in the operating zone before starting machinery.
7. Have your instructor check all special setups and new operations before turning the machine on.
8. Use only power tools with three-prong ground plugs or UL- (Underwriters' Laboratories) approved housings.

Soldering

1. Never test the heat of a soldering iron by feeling it with your hands. Use a piece of solder instead.
2. When not using the soldering iron, always return it to its holder.
3. Remove excess molten solder with the wet sponge on the solder station. Never shake it off onto the bench or floor. Molten solder inflicts painful burns when it touches the skin.
4. Do not pass a soldering iron directly to another individual. Place it back on the iron rest, and let the other person pick it up from there.
5. When performing any soldering operation around electrical cords or any established circuits, be sure that the power is off. Also, be sure that all large capacitors are discharged before soldering any connection.

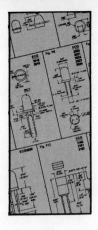

Chapter 2: Basic Concepts

BASIC ELECTRONIC HAND TOOLS

Following are descriptions and illustrations of the basic electronic hand tools used in the assignments and projects covered in this text.

Adjustable Wrench

The adjustable wrench is an open-ended wrench that can be used on nuts and bolts of different sizes because of its adjustable jaw size. It has one stationary jaw and one adjustable jaw that is operated by a thumb screw. To adjust the jaw size, turn the adjustment screw until the jaws are opened wide enough to slip over the nut. Position the jaws over the opposite flats of the nut. Then, turn the screw until the movable jaw grips the flat side of the nut firmly.

Proper Use. When using an adjustable wrench to tighten or loosen nuts, always remember to pull the wrench handle in the direction toward the movable jaw. This will prevent excess force on the movable jaw.

Diagonal Cutting Pliers

Diagonal cutting pliers, also known as diagonals or dykes, are used to cut soft metal wire. They are available in a variety of sizes and styles to accommodate different jobs.

Proper Use. Never use diagonals that are too small for large-gauge wire. Diagonals are not designed to be used as an insulation-stripping tool. Wire nicks may result and cause wire weakening.

When using diagonals, place the cutting edge against the wire or lead to be cut. Avoid using diagonals with dull

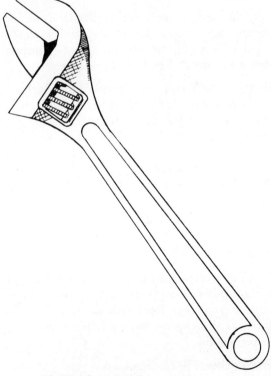

FIGURE 1-3 Adjustable wrench

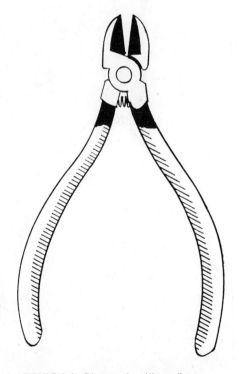

FIGURE 1-4 Diagonal cutting pliers

or nicked cutting edges. For ease of operation, periodically lubricate the joints of diagonals.

Needle Nose Pliers

Needle nose pliers, also called chain nose, or long nose pliers, are available in a variety of sizes, to be used in different applications. They are very versatile and are used in a number of assembly processes. They are used primarily for shaping component or wire leads, crimping wires on terminal posts, and holding hardware during chassis assembly. Most needle nose pliers also have a cutting jaw for cutting small-gauge wires.

Proper Use. Do not use needle nose pliers for any purpose other than those mentioned in the preceding passage. They have very brittle metal jaws that will twist and break if forced. It is important to select the proper size of pliers for the specific job. To insure easy one-handed operation of the pliers, periodically lubricate the joints.

Nut Driver

The nut driver, also called the spin-tite, is designed to rapidly install or remove nuts. It resembles a socket wrench

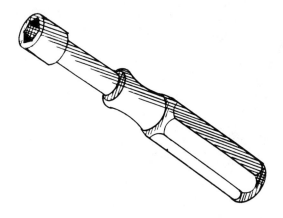

FIGURE 1-6 Nut driver

attached to a screwdriver handle. Nut drivers are available in sizes ranging from 3/32" to 1/2" and can be purchased in color-coded sets. Sets that have long, hollow shafts to allow tightening of nuts onto protruding screws are also available. They are particularly useful in hard-to-reach or congested areas.

Proper Use. When using a nut driver to tighten the nut or screw portion of a hardware assembly, cautiously apply torque pressure. Excessive torque pressure will strip the threads on the screw.

Screwdriver

The screwdriver is used to produce a twisting motion to tighten or loosen slotted screws. The most common types of screwdrivers are the slotted (standard) screwdriver and the Phillips head screwdriver.

Proper Use. To use a screwdriver properly, hold the heel of the handle inside the cup of your hand. Insert the tip of the screwdriver into the screw head, matching the tip to the machine screw head recess. Then, apply a twisting motion.

This tool is not designed to be used as a chisel. Nicked or sharpened tips are dangerous and should be reconditioned or replaced.

Soldering Aid

The soldering aid is a versatile tool that can be used for a number of applications. Its primary function is to remove excess rosin from soldered joints. It is also used to form, or dress, wires around terminal posts, and as an inspection aid.

Proper Use. The blade point is normally used to separate leads or wires when inspecting solder joints. The slotted

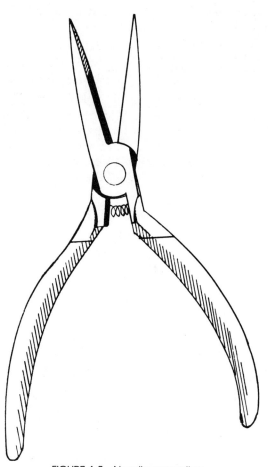

FIGURE 1-5 Needle nose pliers

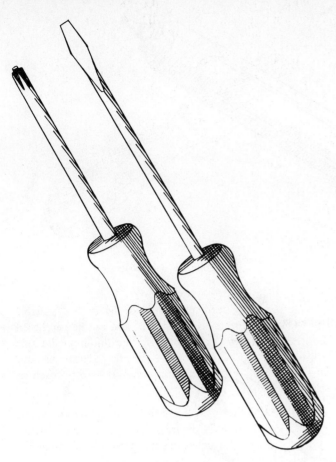

FIGURE 1-7 Screwdrivers

point is designed to form leads around terminal posts before crimp is applied.

Wire Strippers

Wire strippers are used to remove insulation from conductors or wires. They are available in a variety of styles and sizes. The basic wire stripper contains a single slot adjustment within the handle unit. Other wire strippers have cutting edges numbered according to the standard wire gauge number for easy identification.

Proper Use. Proper use of this tool is essential to prevent a bad wire nick or possible damage to the cutting edges. Avoid using dull wire strippers to strip wires.

Solder Sucker

The function of any desoldering device is to remove molten solder from the soldered termination. The desoldering tool illustrated is commonly called a solder extractor or a solder sucker. Most desoldering tools draw the molten solder from the connection with a vacuum or suction force.

Proper Use. A simple cocking action prepares the solder remover for use. The soldered joint is heated as the tip of the desoldering remover is applied to the molten solder. A button or other triggering mechanism is activated causing the molten solder to be drawn into the cylindrical cavity.

Periodically clean and lubricate the solder remover for proper operation. Never place the tip of the solder extractor directly on the soldering iron tip. It is made of a plastic-type material that can be damaged by the intense heat of the tip. Replace the tip when necessary.

SOLDERING BASICS

The basic purpose of soldering is to form a metallurgical bond between metals using a filler metal (solder) that melts below 800 degrees Fahrenheit (427 degrees Celsius). Soldering allows the joining together, both mechanically and electrically, of metal objects such as wires, components, and leads, using a material called solder and a heating device called a soldering iron. The necessary ingredients in the soldering process are illustrated in figure 1-11.

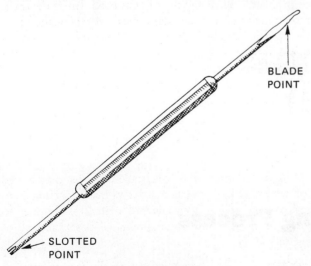

FIGURE 1-8 Soldering aid

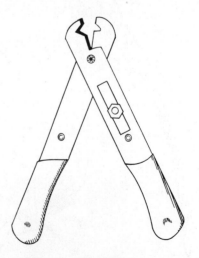

FIGURE 1-9 Wire strippers

Basic Concepts 13

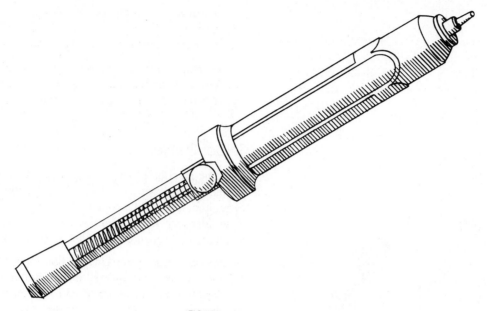

FIGURE 1-10 Solder sucker

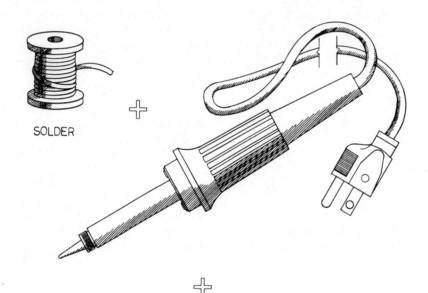

SOLDER   +

+

PRACTICE   =
(Skill)

 **Soldering Process**

FIGURE 1-11 Soldering process

Solder

The most commonly used solder for electronic work is 60/40 rosin-core wire solder. This solder is an alloy composed of approximately 60% tin and 40% lead. Inside the solder, there is a core of semiliquid flux which breaks down the oxide from the surfaces of the metal during the soldering operation (see figure 1-12).

Rosin-core wire solder has a melting point of approximately 370 degrees F (188 degrees C). Below this temperature, but well above room temperature, the rosin flux is an active cleaning agent. As the connection is being soldered, the flux removes minor oxides and film from the metal surfaces and causes them to float to the surface of the connection. After the soldered connection has cooled, the rosin solidifies and returns to an inactive state. Although this change in the rosin does not adversely affect the electrical connection, the solidified rosin should be removed because there may be some conductive particles trapped in the rosin residue.

Industrial-type, chlorinated, flux cleaning agents, such as trichlorethylene, are effective flux-removing solvents. However, because they emit toxic vapors and are flammable, many electronic manufacturing plants discourage their use. Substitutes for chlorinated flux cleaning agents are alcohol compounds such as methanol, ethanol, and isopropyl alcohol, or simple lacquer thinner.

Some industries have discouraged cleaning the flux residue because they have discovered electrical problems created by harmful oxidation developing around connections as a result of the flux removal.

Acid-core solder should *never be used* because of the highly corrosive nature of the acid fluxing agent.

The decision of what diameter of wire solder to use depends on the sizes of the component leads and terminal pads to be soldered. Diameters of 1/16" or 1/32" are the most common sizes used in industry.

The Soldering Iron

Hand-held soldering irons come in a variety of styles and sizes. Industrial-grade soldering irons have replaceable tips, which allow them to be used for specialized applications. For soldering most average-size wire connections, terminals, or printed-circuit boards, a 40-watt to 50-watt iron will give sufficient heat in a short period of time. For heavier work, a larger wattage iron may be needed.

Soldering stations have become extremely popular because of their low-voltage heating elements and temperature control capabilities. For example, the Weller WTCP series, which is recommended for general soldering, has a variety of different-shaped tips in 600-degree, 700-degree, and 800-degree Fahrenheit temperatures (315-degree, 371-degree, and 426.5-degree Celsius) (see figure 1-13).

The selection of soldering iron and tip is extremely important for obtaining quality soldered connections. For

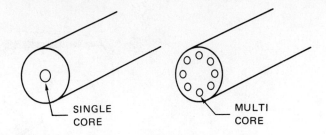

FIGURE 1-12 Cross section of solder wire

general-purpose, printed-circuit board soldering, a 1/16" to 1/8" screwdriver blade or chisel is most commonly used. In figure 1-14 the various tip shapes available for special applications are shown.

Most quality tips are made of copper that has been plated with an ironclad coating. The copper provides an excellent thermal conductive medium, while the ironclad coating provides a hard, protective surface. Although the iron plating reduces the thermal conductivity of the tip, it increases the tip's resistance to pitting which could be a problem if copper were used alone.

Soldering iron manufacturers specify temperature ranges in which different tips can be used. To estimate tip temperature, apply a small amount of solder to the flat surface of the tip. Then wipe the surface with a damp sponge or paper towel. The tip temperature can be estimated by observing the color of the surface immediately after wiping. The tip is *silver* in color at temperatures between *600 degrees F* and *700 degrees F (315 degrees C and 371 degrees C)*. If *gold streaks* appear, the temperature is approaching *800 degrees F (427 degrees C)*.

The procedure for keeping tips clean during the soldering operation depends on the type of tip being used. The solid copper tip should be coated (*tinned*) with solder when used for the first time, and periodically thereafter to insure a clean and efficient heat transfer surface. When solder is added to the tip, and also during the soldering operation, the flux in a rosin-core solder will help keep the faces on the end of the tip tinned. When this type of tip develops a rough or pitted surface, it may be necessary to file the pitted areas and recoat them with solder.

To care for the ironclad, plated tip during its use, continually wipe it with a wet rag or sponge while tinning the faces with a coat of solder. *Never file* a plated tip. This will destroy the plating and eliminate the advantages of this type of tip (see figure 1-15).

The Soldering Process

To begin the soldering process, the connections or components to be soldered must be prepared. Wires on terminals must be wrapped (*crimped*) to specifications, or components on printed-circuit boards must be loaded and ready for soldering.

After the terminal or printed-circuit board is prepared, the solder is applied to make the connection complete. The

Basic Concepts 15

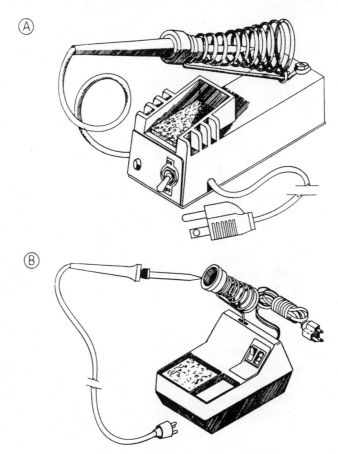

FIGURE 1-13 Weller soldering stations

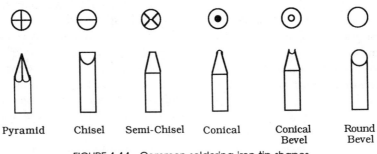

FIGURE 1-14 Common soldering iron tip shapes

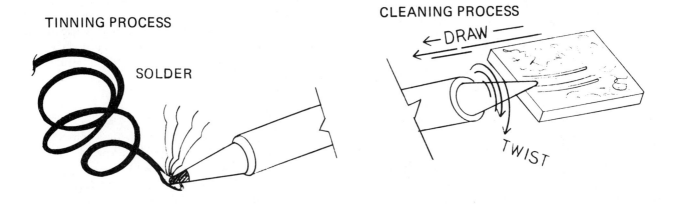

FIGURE 1-15 Tip tinning and cleaning process

16 Introduction

solder can be applied using a hand-held soldering iron, resistance soldering, dip soldering, percolation, or induction soldering. The method chosen is adapted to meet individual requirements, and in many cases, a production-line situation. The hand-held soldering iron is still an important tool application in small production or repair work.

During the actual soldering operation, care must be exercised to prevent such undesirable results as the cold solder joint, the fractured joint, the rosin joint, and wicking or solder creep. If the component leads to be soldered have not been heated enough, a *cold solder joint* will result. In this type of undesirable soldered connection, the solder does not melt sufficiently to flow smoothly.

The *fractured joint* is one in which some sort of movement causes the parts of the joint to move while the molten solder is in its plastic state. In the plastic state, the solder is neither solid nor liquid, and any movement causes it to crystallize, separate, and form a very rough connection.

A *rosin joint* is one in which a part of the joint has been heated enough to become coated with flux, but not to melt the solder. This part of the joint is left coated with rosin flux, which acts as an insulator and consequently provides a poor electrical connection or no electrical connection.

Wicking or *solder creep* is an undesirable effect when soldering stranded wire. Since solder will be drawn to heat, if a terminal or wire is heated too much, the solder will follow the heat back under the insulation of the wire and cause the wire to become a solid strand. Any vibration may then cause the wire to break at the connection point due to the rigid stranded wire created by solder creep. Excessive heat may also cause the insulation to burn or melt.

To avoid these soldering problems, it is extremely important to work with a clean soldering iron and clean workpieces. It is also necessary to have sufficient heat on the parts being soldered, but the heat must not be allowed to damage the components or wires involved. These problems can be avoided by supplying sufficient heat to the work quickly, and completing the operation in a short period of time. The heat will transfer more effectively from the soldering iron tip to the work if the tip is clean and tinned so that a bridge of solder can be formed at the point of contact. Sometimes a small amount of solder may need to be added to form this bridge. This technique is illustrated in figure 1-16.

Holding the Soldering Iron

The proper method of holding the soldering iron varies from one individual to the next. The grip used around the handle of the soldering iron is identical to that used when holding a pencil or any writing apparatus. Regardless of the style of grip used, the important consideration is the comfort one develops with the "natural" method acquired from experience. The quality of the soldering job can be directly related to the level of skill in manipulating this important tool to the workpiece.

The position of the soldering tip is another factor governing the quality of soldered connections. The technique used to apply heat to the terminal while feeding solder to the heated surface is determined by the type of soldered connection desired. For example, when soldering component leads to a printed-circuit board, the soldering tip must first make contact with *both* the copper foil and the component lead. The solder is then fed to the heated connection, not to the soldering iron tip. When soldering components or wires to any type of terminal lugs, heat must be applied to the component lead or leads crimped to the terminal lug before solder is applied to the heated junction.

Although learning the rules is important in developing soldering technique, soldering skill is acquired through practice. In the assignments and projects in Sections 3 and 4, this skill is practiced.

CONDUCTORS AND CABLES

There are two basic methods of connecting electrical circuits: by conductors in the form of braided wire, insulated wire, or cable; or by waveguide conductors made of aluminum, brass, copper, silver, or other metals. The type of conductor used in a connection is determined primarily by the electrical and mechanical situation the electronic instrumentation or devices are subjected to. For example, conductors used in alternating current sources are generally twisted together in pairs to prevent the possi-

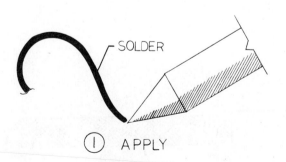
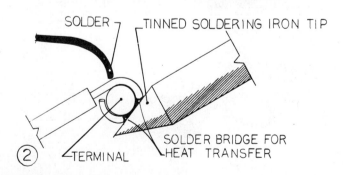

FIGURE 1-16 Solder bridge for heat transfer

bility of ac hum interference from adjacent wires. Conductors are likely to radiate electromagnetic energy that might affect other conductors that have an external shield, which consists of a flexible copper braid woven over the insulation. The shield may or may not have a protective insulation jacket. The shield is normally connected electrically to the metallic chassis.

Cables are usually made of two or more conductors. An exception to this rule is the RF (radio-frequency) coaxial cable, which has a center conductor that is insulated by a non-conductive material, and surrounded by a braided shield along the entire length of the cable (see figure 1-17). Coaxial cables are used in RF circuits as the transmission lines between the antenna of transmission or receiving equipment.

The center conductor in the coaxial cable is centered within the outer shield, either by means of special insulating washers or by insulating material, such as polyethylene or Teflon, that completely fills the space. The size of the center conductor is determined by the current it is to carry. The *characteristic impedance* of the cable is determined by the diameter of the center conductor, the area of the outer shield, and the distance between the center conductor and the outer shield.

Conventional cables may have numerous variations. For instance, they may be made of various amounts of individually insulated conductors, or individually shielded conductors. An example of this type of cable is the common TSP (twisted-shielded pair), shown in figure 1-17.

A system of individual conductor identification is needed in complex electronic wiring assemblies to simplify circuit testing and tracing of individual circuits. Such a system is important in harness assemblies because individual conductors can no longer be traced by eye. Therefore, there must be some system to identify the individual conductors connected in sub-assemblies or components.

Although conductor coding systems vary, they may include such items as: conductor color or colors; AWG (American Wire Gage) sizes; solid or stranded; conductor insulation designated by name, abbreviation, or drawing number; and circuit designation by color or code. An acceptable standard adopted by governmental agencies of the Department of Defense covers details of cables, chemical processes, components, drawings and engineering data, semiconductors, general electronic equipment design, hardware, insulation, instruction handbooks, insulating compounds, materials, and paints—in brief, any item included in the design and construction of electronic equipment and devices.

Other standard organizations have been formed to unify common industrial practices. For example, EIA (Electronic Industries Association) is a national association of electronic manufacturers. It issues engineering standards helpful in producing interchangeable components within the electronic industries. Another common standard organization is the IEEE (Institute of Electrical and Electronic Engineers). The primary objective of this organization is to maintain some consistency and communication between the component manufacturers and the industries that use the components.

INTRODUCTION TO PRINTED-CIRCUIT BOARDS AND SOLDERING

A printed-circuit board (PCB) consists of a layer of insulated, plastic-base material called the *substrate*, supported by a thin, conductive layer of copper foil. Common substrate materials include paper, glass fibers, or interwoven glass mats combined with a bonding resin of phenolic, polyester, or epoxy. A process called *high-pressure lamination*, which unites material under heat and high pressure, is used to form the substrate layer, commonly called the *laminate*. The substrate laminate provides the

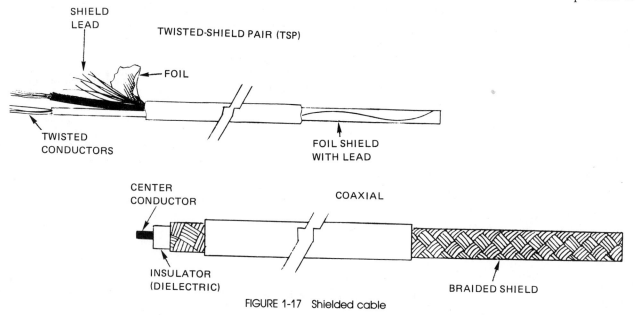

FIGURE 1-17 Shielded cable

foundation to support electrical components as well as the thin layer of copper foil. The copper foil is also laminated using a similar process of heat and pressure to form what is referred to as the *copper-clad laminate*. The copper-clad lamination process is illustrated in figure 1-18.

Copper-clad laminates are manufactured as blank boards on which the actual circuit conductive patterns are applied initially by a layout procedure performed by electronic drafters. One of the fastest and simplest layout procedures is the *direct layout* method. In this method special printed-circuit layout patterns, such as donut pads, transistor patterns, and integrated circuit patterns, and layout tape are applied directly to the copper-clad board. Using the schematic diagram as a reference, the electrical paths are applied to the copper-clad board with the special layout patterns and the drafting tape which serves as the electrical conductor. After the circuit pattern has been applied to the copper-clad board, the board is agitated in an etching solution, or *etchant*, which dissolves the unwanted copper areas around the circuit patterns formed by the tape-up. The board is then removed from the etchant, rinsed, inspected, drilled, and assembled.

Another layout technique known as the *photographic-mask layout* has become a more acceptable production method. This method involves several more intermediate steps than the direct pattern method. Initially, the layout from the schematic diagram is designed on a larger scale using special layout patterns and drafting tape similar to those used in the direct pattern method. This layout is known as the *artwork master*. Utilizing the photographic reduction process, the artwork master is then reduced to the actual finished size of the printed-circuit board. The photo master, or *mask*, which is either a positive or negative film, is then placed on top of a photosensitized, copper-laminated board and exposed to light. After the board has been exposed to light, it is immersed in a developing solution until the layout pattern is visible. The procedures that follow the immersion of the board are similar to the final procedures used in the direct method where the board is etched, drilled, and assembled.

Only a brief description of the printed-circuit fabrication process has been presented in this section. More sophisticated systems of PCB design and development have filtered into the manufacturing process, making this one of the more interesting and challenging areas of the entire manufacturing process. These systems are covered in detail in Sections 3 and 4.

By definition, printed-circuit is synonymous to printed-wiring. Unlike the direct point-to-point wiring method of connecting wire to components, the printed-wire method simplifies the wire termination to component leads. This method of wiring has created another important skill development area that includes the technique of PCB assembly (component mounting), the soldering and desoldering process and circuit board repair. Each of these specialized skill areas is covered in this text.

SCHEMATIC DIAGRAMS AND COMPONENT IDENTIFICATION

Many types of diagrams are necessary to completely describe the construction and operation of electronic equipment. The most important to the service technician is the *schematic diagram*. A schematic is usually all that is required for analyzing, explaining, and servicing most circuits. A schematic, however, cannot convey all of the information about a piece of equipment. Other types of diagrams such as block diagrams, wiring diagrams, printed-

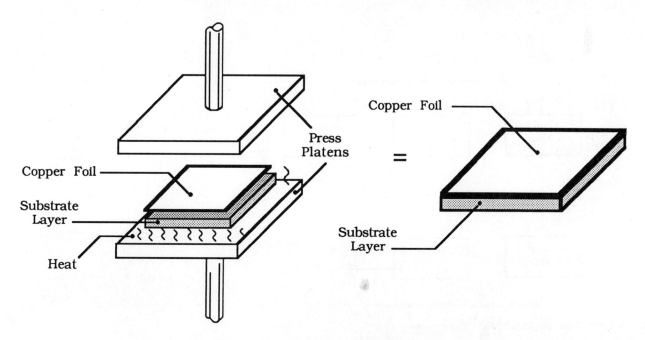

FIGURE 1-18 Copper-clad lamination process

circuit layout diagrams, chassis-layout diagrams, photographic and pictorial diagrams, and mechanical diagrams are needed as well.

Schematic Diagrams

A schematic diagram is a condensation of electrical circuit data represented by accepted graphic symbols drawn in an orderly manner. The electrical connections between components are represented by lines drawn to appropriate points on the symbol. The schematic diagram is usually the first step in the construction of an electronic project since it presents and lists the components that are to be included in the equipment. Later, it is used as the master drawing for the identification and rating of the various components on the chassis-layout diagram, the wiring diagram, or the printed-circuit layout diagram.

The process of laying out a schematic diagram based on an engineering sketch consists of:

1. collecting data about the components on the diagram;
2. organizing the general layout of the diagram;
3. identifying the components by reference designations and other data;
4. adding adequate notes to the drawing; and
5. making the diagram conform in size and detail to other drawings that will be produced for mechanical assemblies.

Most schematics follow the same general arrangement. The input is normally located in the upper left-hand corner of the diagram. From this point, the adjoining circuits are usually arranged in rows from left to right and from top to bottom. The block diagram for an AM radio receiver is illustrated in figure 1-19. Beginning at the input, you can trace your way through the individual circuits as if you were reading a book. The best way to read a schematic is to analyze each stage by the block diagram approach. That is, each block can be isolated by its input and output which feeds the succeeding input of the adjacent block. If this pattern is followed for any schematic, the operation of any equipment—from simple to complex—should become apparent.

For the beginning student, however, analysis of circuit behavior from a schematic diagram may not be as simple as one would like to believe. It is necessary, therefore, to develop an understanding of the relationship between the symbolic circuit diagram and the actual component level. Before actually reading a schematic diagram, the letters that have been assigned to the various electronic components and the miscellaneous electronic support items must be learned (or memorized). These letters are known as reference designations. After the reference designations are learned, the schematic symbols and the physical components which they represent must be learned. Finally, these symbols (components) can be connected to make up a very simple schematic diagram.

Reference Designations

The letter and number combination beside each component symbol is called the *reference designation*. The letter signifies the type of component and the number distinguishes the component from all others of the same type.

Reference designations can be classified according to two general categories—those used for consumer or relatively simple equipment, and those used for military and complex equipment. In general, military equipment is

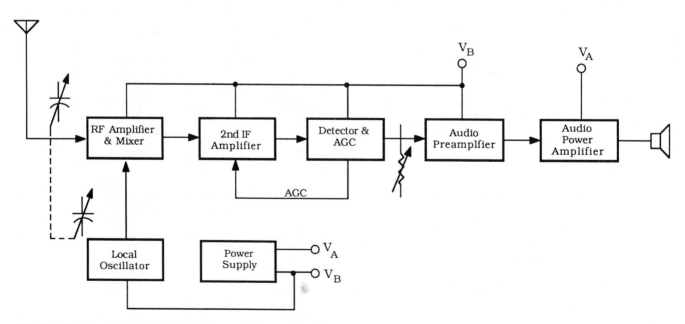

FIGURE 1-19 Block diagram of AM radio receiver

20 Introduction

more complex than equipment made for consumer use; therefore, it is necessary to use different methods of component identification for military equipment and consumer equipment. In many cases, however, identical practices are followed for both.

Unfortunately, not all electronic manufacturers designate a given component by the same letter code. However, the designations are fairly standard, and usually only a few items will be different. The recommended reference designations are shown in table 1-4.

TABLE 1-4 Reference Designations for Consumer and Military Equipment

Item	Code	Item	Code	Item	Code	Item	Code
adapter, connector	CP	coupling (aperture, loop, or probe)	CP	inductor	L	phase changing network	Z
amplifier	AR			inseparable assembly	U	photodiode	CR
amplifier, magnetic (except rotating)	AR	crystal unit, piezoelectric; crystal unit, quartz	Y	instrument	M	phototransistor (isolator)	A,U
amplifier, operational	AR			insulator	E	phototube, photoelectric cell	V
amplifier, rotating (regulating generator)	G	current regulator (semiconductor device)	CR	integrated circuit package	U		
				integrator	A	plug, electrical (connector, movable portion)	P
amplifier, summing	AR	cutout, fuse	F	interlock, mechanical	MP		
antenna	E	delay function	DL	interlock, safety, electrical	S	potentiometer	R
antenna, radar	E	delay line	DL				
arrester, lightning	E	diode, breakdown	VR	inverter, motor-generator	MG	power supply	PS
assembly, inseparable or nonrepairable	U	diode, capacitive	CR	jack	J	reactor	L
		diode, semiconductor	CR	junction (coaxial or waveguide)	CP	receiver, radio	RE
assembly, separable or repairable	A	diode, storage	CR			receiver, telephone	HT
		diode, tunnel	CR	junction, hybrid (magic tee)	HY	receptacle (connector, stationary portion)	J
attenuator (fixed or variable)	AT	disconnecting device (connector, receptacle)	J	lamp	DS	recorder, elapsed time	M
audible signalling device	DS	disconnecting device (connector, plug)	P	lamp, fluorescent	DS	recording unit	A
autotransformer	T			lamp, glow	DS	rectifier (semiconductor device, diode)	CR
backward diode	D,CR	disconnecting device (switch)	S	lamp, incandescent	DS		
ballast tube or lamp	RT			lamp, pilot	DS	rectifier, semiconductor controlled	Q
barrier photocell	V	discontinuity (usually coaxial or waveguide transmission)	Z	lamp, resistance	RT		
battery	BT			lamp, signal	DS	rectifier (complete power supply assembly)	PS
block, connecting	TB			lampholder	X		
blocking layer cell	V	divider, electronic	A	light emitting diode	DS	regulator, voltage	V
blower	B	dynamotor	MG	line, artificial	Z	relay	K
brush, electrical contact	E	electronic divider	A	loop antenna	E	resistor	R
bus bar	W	electronic function generator	A	loudspeaker	LS	resistor, current regulating	RT
cable, cable assembly (with connectors)	W			mechanical part	MP		
		electronic multiplier	A	meter	M	resistor, terminating	AT
capacitor bushing	C	equalizer; network, equalizing	EQ	microcircuit	U	resistor, thermal	RT
capacitive diode	CR			micromodule	U	resistor, voltage sensitive	RV
capacitor	C	facsimile set	A	microphone	MK	resolver	B
cavity, tuned	Z	fan; centrifugal fan	B	mode suppressor	Z	resonator (tuned cavity)	Z
cell, battery	BT	ferrite bead rings	E	mode transducer	MT	rheostat	R
cell, solar	BT	field effect transistor	Q	modulator	A	rotary joint (microwave)	E
choke coil	L	filter	FL	motor	B	semiconductor controlled switch	Q
chopper, electronic	G	fuse	F	motor-generator	MG		
circuit breaker	CB	fuseholder	X	multiplier, electronic	A	semiconductor controlled rectifier	Q
coil, radio frequency	L	gap (horn, protective, or sphere)	E	network, equalizing	HY		
coil (all not classified as transformers)	L			network, general (where specific class letters do not fit)	Z	semiconductor device, diode	CR
		generator	G			sensor (transducer to electrical power)	A
computer	A	Hall element	E	network, phase changing	Z		
connector, receptacle, electrical	J	handset	HS	oscillator, magnetostriction	Y	servomechanism, positional	A
contact, electrical	E	hardware (common fasteners, etc.)	H	oscillograph	M	shield, electrical	E
contactor (manually, mechanically, or thermally operated)	S	head (with various modifiers)	PU	oscilloscope	M	shield, optical	E
				pad	AT	shifter, phase	Z
contactor, magnetically operated	K	headset, electrical	HT	part, miscellaneous electrical	E	shunt, instrument	R
		heater	HR			signal light	DS
core, air; magnet; magnetic; storage	E	horn, electrical	LS	part, miscellaneous mechanical (bearing, coupling, gear, shaft, etc.)	MP	slip ring (ring, electrical contact)	SR
		hydraulic part	HP			socket	X
coupler, directional	DC	indicator (except meter or thermometer)	DS			solenoid, electrical	L

TABLE 1-4 Continued

speaker	LS	switch, semiconductor		transmitter, radio	TR	waveguide	W
squib, electric	SQ	controlled	Q	triode, thyristor	Q	waveguide flange	
squib, explosive	SQ	taper, coaxial or		triac, gated switch	Q	(choke)	J
stabistor	CR	waveguide	T	tuner, E-H	Z	waveguide flange (plain)	P
strip, terminal	TB	teleprinter	A	varactor	D,CR	winding	L
subassembly, separable		transformer	T	varistor, asymmetrical	D,CR		
or repairable	A	transistor	Q	varistor, symmetrical	RV		
subdivision, equipment	N	transmission line,		vibrator, interruptor	G		
switch	S	strip-type	W	voltage regulator			
switch, interlock	S	transmission path	W	(semiconductor device)	VR		

Schematic Symbols of Components

As with the reference designations, the schematic symbols used by various electronic manufacturers differ from one company to the next. The symbol for each component is discussed in this text, and where differences exist, the various ways of depicting a given item are shown.

Efforts have been made by organizations throughout the world to standardize the symbols used by industries. The IEEE, the American National Standards Institute (ANSI), and the military services (MIL Spec) have adopted standard symbols which they hope industry will use. The Electronic Industries Association (EIA) has been instrumental in coordinating the efforts of various groups to standardize schematic symbols. And, the International Electrotechnical Commission (IEC) has recommended standard symbols (with which most ANSI and IEEE symbols agree) to member countries throughout the world.

It is interesting to note that as a result of the efforts of the IEC, the differences between schematic diagrams for products manufactured by two American companies may be greater than the differences between the schematic diagrams for products manufactured by U.S. and foreign companies.

Fortunately, the differences between the symbols used by various companies to identify a component are not as great as in previous years. Today, most of the differences between the symbols stem from the differences in drafting practices and in methods of circuit layout. Schematics may be hand sketched, drawn with standard templates and letter guides, produced by pre-printed symbols, or even prepared by a special typesetting machine. Therefore, minor differences are inevitable. Some of the obvious differences, however, have no effect on the electrical interpretation of the drawing.

Graphic symbols are formed from such simple elements as arcs, angles, circles, rectangles, semicircles, short lines, squares, tees, triangles, and zigzag lines. The meaning of a symbol is not altered by its orientation. However, by convention, many symbols are almost always drawn in one particular direction. An example of this is the ground symbol which is illustrated in Section 2.

Although heavier lines are sometimes used for emphasis, line width on symbols does not alter the meaning of the symbol.

Symbol Modifiers

Symbol modifiers are used to modify some of the basic graphic symbols. Examples of symbol modifiers are shown in figure 1-20. The three *adjustable* modifiers shown—the adjustable, linear, and nonlinear modifiers—should be drawn at a 45-degree angle *across* the body of the basic symbol.

The well-known polarity symbols, + and −, are used to indicate the polarity of a source or component. An example of the use of polarity symbols is seen on the symbol for an electrolytic capacitor, or on the positive and negative terminals of a dc voltage source.

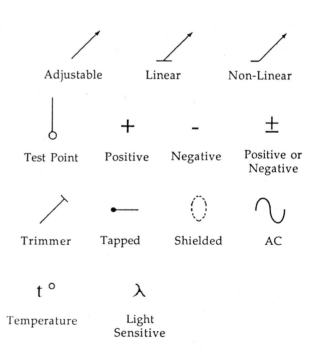

FIGURE 1-20 Symbol modifiers

Mechanical Functions. Another type of symbol modifier is used to identify such mechanical functions as linkage or motion on circuit diagrams. The symbol for mechanical connection or linkage is shown in figure 1-21. It consists of a series of dashes that connect such linked mechanical devices as switch sections, which are separated on the schematic.

GANGED POTENTIOMETER

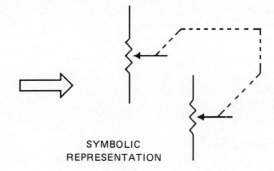

SYMBOLIC REPRESENTATION

FIGURE 1-21 Mechanical function modifiers

SECTION 2
ELECTRONIC COMPONENT SYMBOLS

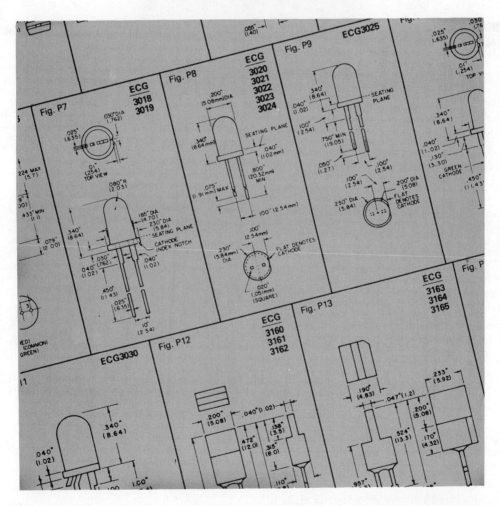

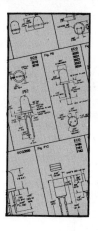

Chapter 3: Graphic Symbols of Common Electronic Components

Each of the graphic symbols contained in the preceding reference guide is described briefly in this chapter. The descriptions include explanations of the physical and/or electrical properties of each component.

The intent of this chapter is to provide the student with basic information which will enable him/her to identify different components. For this reason, the descriptions are not highly detailed.

UNIT 1 BASIC PASSIVE COMPONENTS

RESISTORS

The resistor is probably the most common of all electronic components. In electronic circuits, the resistor functions exactly as its name implies—it resists, or opposes, current flow. Every electrical component or conductor contains a certain amount of resistance. By using resistors, practically any amount of resistance may be introduced into electronic circuitry in a small compact unit.

Resistors can be identified by two electrical quantities. The first is the electrical value given in *ohms*, the unit of resistance. By proper selection of resistance value, it is possible to obtain an exact amount of current flow in certain parts of a circuit. The second value given in rating a resistor is the wattage. The unit of measurement is the *watt*. The wattage rating of any resistor is its ability to dissipate heat as current flows through its material composition. Generally, the larger the size of the resistor, the higher its wattage rating. Common ratings are 1/4 watt, 1/2 watt, 1 watt, 2 watts, and 5 watts or more.

Fixed Resistors

The fixed resistor is the most commonly used resistor. As the word fixed implies, it is constructed in such a way that its ohmic value cannot be varied. The ohmic value is controlled by the amount and type of material from which the resistor is made. Fixed resistors are normally made of carbon. Some examples of fixed carbon resistors are shown in figure 2-1. The reference designation assigned to all resistors is the letter *R*.

Fixed carbon resistors are usually marked in some way to indicate their ohmic value. This is often done by a system of color coding, which is outlined in the Appendix for both four and five band systems.

The schematic symbols for fixed resistors are illustrated in figure 2-2. Symbol A is the most common in the United States while symbol B is a popular symbol in foreign schematics.

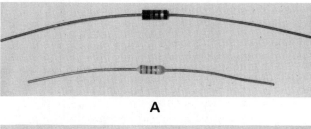

A

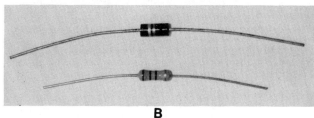

B

FIGURE 2-1 Carbon resistors: (A) ¼-watt, 5% tolerance, (B) ½-watt, 10% tolerance

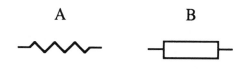

FIGURE 2-2 Fixed resistor symbols

On other types of fixed resistors, especially those intended for high power (wattage) applications, a special high-resistance wire is wound on a ceramic core, or another insulating core. This assembly is then covered with a vitreous-enamel protective coating. A cutaway view of this type of fixed resistor is shown in figure 2-3. Normally, the value and wattage rating is stamped on this resistor's side.

Tapped and Adjustable Resistors

A fixed resistor may have a tap or connection at some point along the resistance material. The symbols used for such a resistor are shown in figure 2-4. Symbol A is used almost exclusively. As you may have noticed, it is identical to the symbol for the fixed resistor except for the line connected to the zigzag portion of the symbol. If there are two taps, another line is added as shown in B. A dot is sometimes added at the point where the line connects the resistor symbol as shown in C.

A high-wattage tapped resistor is shown in figure 2-5.

Variable Resistors

Variable resistors can be continuously varied in value as required. In high-power applications, a variable resistor is usually called a *rheostat*. A rheostat consists of a wirewound resistance element arranged in a circle. A sliding contact, which is electrically connected to one end of the resistance element, enables the total resistance to be continuously varied from the maximum rated resistance to minimum resistance. Rheostats are normally used to limit current in high-power applications.

Another type of continuously variable resistor is the *potentiometer*, or *pot*. In appearance, it is very similar to the rheostat. It can be made of wirewound material, carbon, or other resistance material. The pot has connections to each end of the total resistance element, and a sliding arm (wiper).

If the wiper is connected to one of the outside terminals, the pot can function as a rheostat. The potentiometer and rheostat, therefore, are very similar variable resistance components. However, a rheostat can only be converted to a pot if its adjustable arm can be electrically disconnected from the outside resistance element. The basic rheostat can be identified as a two-terminal device, while the potentiometer is a three-terminal device.

The electrical functions of these two versatile variable resistors determine their circuit applications. The rheostat is a *current-limiting device* while the potentiometer is a *variable voltage divider*. The graphic symbols for rheostats

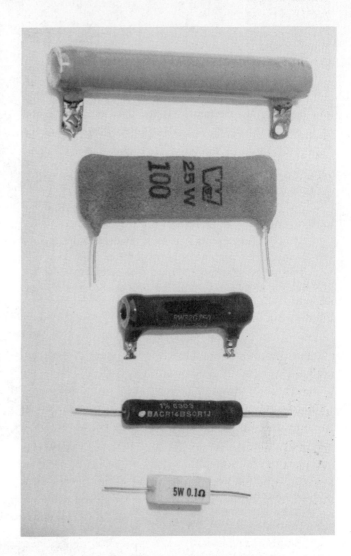

FIGURE 2-3 Wirewound fixed resistors

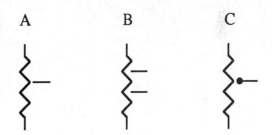

FIGURE 2-4 Tapped resistor symbols

FIGURE 2-5 Tapped resistors

Graphic Symbols of Common Electronic Components

and potentiometers and illustrations of these devices are shown in figure 2-6.

The most common potentiometer symbols are A and B. They are alike except that *no connection* is shown at one end of symbol B. This does not necessarily mean that this unit has only two terminals. If the third terminal is not used, as is often the case, it is not shown as being connected.

Symbols C and D also designate a rheostat or potentiometer with only two terminals in use. The arrow as the basic symbol modifier signifies that the device is adjustable. Two other symbols that show the variable arm are E and F.

Symbol G represents a preset screwdriver adjustment.

Symbol H represents a potentiometer that has a tap which provides an additional fixed connection. Some units are made in such a way that the sliding contact will not move beyond a certain point on the element. This will always leave some resistance between the arm and one end of the potentiometer.

The value of a variable resistor may not always be stamped on the unit. The manufacturer's part number is often the only information given. To find the correct value, refer to the schematic, the parts list, or the manufacturer's catalog.

The most popular reference designation for variable resistance is the letter *R*. Some schematics, however, use the letter *P* for potentiometer, or *VR* for variable resistor.

Special Resistors

Often there is a need for a special type of resistor which will vary in value when the environmental (tempera-

RHEOSTATS

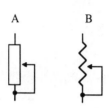

POTENTIOMETERS

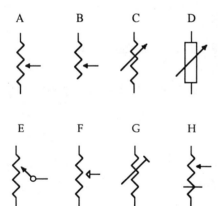

TRIMMERS

FIGURE 2-6 Rheostats and potentiometers

ture) conditions change. Many of these special resistors are semiconductor devices known as *thermistors*. Since they are used in circuits as resistance components, the symbol used to identify them is the basic resistance symbol with the appropriate modifier attached to represent the specific type of resistor (see figure 2-7).

Another special type of resistor functions as a fuse. Its circuit application is similar to that of any fuse: if the current through it exceeds a certain value, it opens to protect expensive circuit components or entire circuits. Such units are often made to plug into a socket on a chassis, making replacement a simple task. The graphic symbols that represent these special fusible resistors are illustrated in figure 2-8.

If the resistance of these special resistors increases as the surrounding temperature rises, they are said to have a *positive temperature coefficient (PTC)*. If their value decreases as the temperature rises, they have a *negative temperature coefficient (NTC)*. The abbreviation signifying the specific type is usually placed beside the symbol.

Resistors can also be made to change in value according to the current flowing through them or the voltage present in the circuit. Others vary in value when light strikes their surfaces. The symbol that represents these special resistors is usually the basic resistor symbol with a letter designating the type of change; for example *V* for voltage, *I* for current, and *L* for light. Other letter abbreviations are *RT* for temperature-dependent resistor, *RV* for voltage-dependent resistor, and *LDR* for light-dependent resistor. Some of these special resistor symbols are illustrated in figure 2-9.

CAPACITORS

The capacitor, like the resistor, is found in nearly all electronic equipment. A capacitor is two metallic plates separated by an insulating material called a *dielectric*. The physical construction of this component makes it able to store (*charge*) and release (*discharge*) electrons as dictated by the external electrical conditions affecting it.

The property by which a capacitor is able to store electrons is called *capacitance*. Capacitance is directly affected by the size of the plates; the larger the area of the plates, the more electrons the capacitor is able to store, and the larger the capacitance. Capacitance is also affected by the distance between the plates and the material used as the dielectric.

The unit of capacitance value is the *farad (F)*. Since the farad represents a very large electrical unit, the *microfarad* (μF) and the *picofarad* (pF or μμF) are used more often as units of capacitance.

Capacitors are also rated by their *breakdown voltage* (in volts dc), and their *tolerance* of their rated capacitance value (in percent).

Capacitors can be *fixed, variable,* or *adjustable* types.

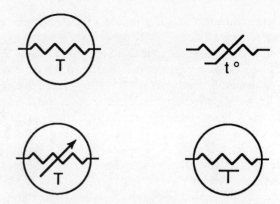

FIGURE 2-7 Temperature-compensating resistor symbols

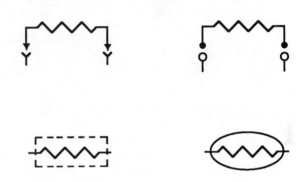

FIGURE 2-8 Fusible resistor symbols

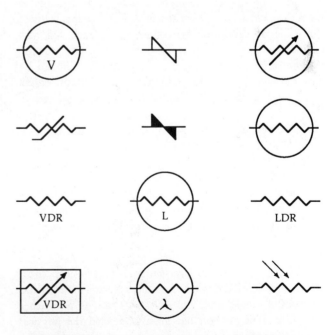

FIGURE 2-9 Voltage, current, and light-dependent resistor symbols

Graphic Symbols of Common Electronic Components 29

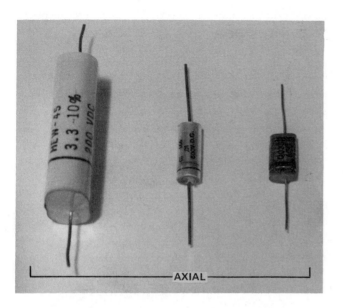
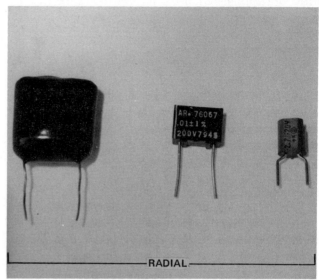

FIGURE 2-10 Fixed axial and radial paper capacitors

Fixed Capacitors

As their name implies, fixed capacitors have capacitances that cannot be adjusted. One of the most widely used fixed capacitors is the *paper-foil* type. It contains two metallic, foil strips separated by two or more layers of very thin paper which have been impregnated with wax, mineral oil, or some other dielectric material. The number of paper layers determines the voltage breakdown value of the capacitor. This whole assembly is rolled into a cylinder to form the axial-style, paper capacitor. The terminal leads of this type of capacitor are connected to each foil that extends beyond the paper. This method of construction eliminates inductance, which is the property exhibited by any conductor in a coil form. On the other hand, the radial-style, fixed capacitor lead terminations are connected to each foil.

The cylindrical construction of the capacitor makes it possible to enclose the assembly in a cardboard, plastic, or ceramic tube with radial leads projecting through the end seals. This *tubular* capacitor is marked with a black band on the end where the lead is connected to the outer foil. This lead is normally connected to the ground potential to act as a shield. Axial-style and radial-style capacitors are shown in figure 2-10.

A variation of the basic paper capacitor is the *metallized-paper* or *plastic capacitor*. In this type of capacitor, the aluminum foil is deposited on the dielectric material and becomes a single unit, but still performs as any other capacitor. This method has greatly reduced the physical size of the capacitor without varying from the ratings of a comparable paper capacitor.

The *mica capacitor* is another type of fixed capacitor. In this capacitor, the plates are arranged in layers separated by thin sheets of the dielectric material, mica. The assembly is then covered with a durable plastic material to protect it from external contamination. This type of capacitor is normally found in the radial style as shown in figure 2-11.

Another type of mica capacitor not commonly used in modern electronic equipment is also illustrated in figure 2-11. This is a pure mica capacitor. It was used before the silver-mica alloy was developed for use as a dielectric material. Dots are used to color code this capacitor; an

FIGURE 2-11 Molded mica capacitors

30 Electronic Component Symbols

arrow or similar indicator on the dots shows the reading sequence. Colored dots at specific positions in the reading system may indicate military or commercial usage, temperature coefficient, or tolerance.

The *ceramic fixed capacitor* is constructed very much like the mica capacitor, except that when ceramic material is used as the dielectric material, silver coating is used to make the conductive plates. The silver coating is deposited on opposite sides of the ceramic material and is connected to the terminals at the ends. These terminals are then connected to leads. Ceramic capacitors are constructed in tubular, flat disc, and other shapes. A construction view of a ceramic disc capacitor is shown in figure 2-12.

Plastic films such as polyester, polystyrene, polycarbonate, polytetrafluoroethylene, and polypropylene are used as dielectric materials in other types of capacitors. All capacitors, however, have the same basic parts—a conductive surface separated by an insulator (dielectric).

Usually, the capacitance value of a fixed capacitor is stamped on its body, or a color system gives its value and other information, such as the *working voltage*. The working voltage is the amount of dc voltage that can be continuously applied across the capacitor without causing the dielectric material to break down, resulting in arcing between the plates.

Symbols for fixed capacitors are shown in figure 2-13. Symbol A is used most often on diagrams, having superseded symbol B which was used for many years. Symbols C through F represent *feedthrough capacitors* which are also illustrated in figure 2-13. Feedthrough capacitors are used to couple electrical signals from encased subassemblies to required destinations such as other subassemblies or external terminal points of the main chassis unit.

Electrolytic Capacitors. The *electrolytic capacitor* is a type of fixed capacitor characterized by a large value of capacitance in comparison to its size. Electrolytic capacitors have one plate made of a moist conducting substance called an *electrolyte*. Some metals such as aluminum and tantalum have the ability to form an oxide layer when immersed in an electrolyte solution. This oxide film becomes the insulation or dielectric layer between the metal plate and the electrolyte which serves as the other plate.

Unlike with other capacitors, the *polarity* of the terminal leads must be considered when connecting the electrolytic capacitor in any circuit. The positive lead must be connected to the point with the *most positive* voltage, and the negative lead must be connected to the *most negative* potential.

Paper, mica, ceramic, and the other types of fixed capacitors previously discussed seldom have a value greater than one microfarad, whereas electrolytics range from a tenth of a microfarad to one farad. Because of their large values, electrolytics can store many more electrons. This makes them useful for *filtering*—smoothing out variations

FEEDTHROUGH CAPACITORS

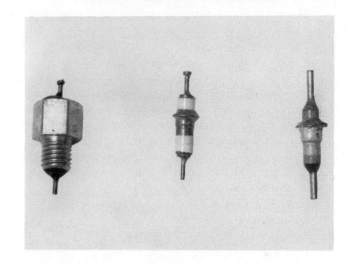

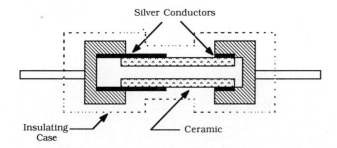

FIGURE 2-12 Ceramic capacitor construction

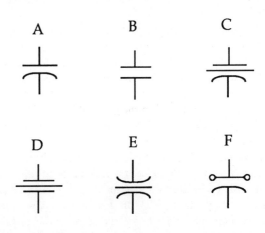

FIGURE 2-13 Fixed capacitor symbols and feedthrough capacitors

in voltage. They are, therefore, used as primary filtering components in power supplies, where they convert ac voltage to a lower or higher value of *dc voltage*. Electrolytics are also used widely as coupling components in audio application circuits.

In applications, however, where a large capacitor is required in a circuit in which the voltage does not change polarity, *nonpolarized capacitors* are used.

The various types of electrolytic packages are illustrated in figure 2-14.

The same symbol used for fixed capacitors is used for electrolytic capacitors, with the addition of a plus (+) sign, and sometimes a minus (−) sign to indicate the polarity of the unit. The various electrolytic capacitor symbols are illustrated in figure 2-15.

Symbol A is used most often on modern schematic diagrams. Symbol B removes any doubt that the capacitor is electrolytic. Other variations of electrolytic capacitor symbols are illustrated at C, D, and E.

More than one electrolytic capacitor is often enclosed in the same container. The negative sides of these units are normally connected together and electrically tied to the "can." Separate terminals or leads, however, are provided for the positive sides of each capacitor. These multisection capacitors are sometimes designated by symbol F. Notice the small *rectangle* and *triangle* near the two sections in the symbol. These figures are stamped on the sides of the can along with the respective value ratings of each section. Thus, they serve to identify the separate sections.

Symbol G is the preferred symbol to indicate a two-section capacitor. Symbol H is used to identify nonpolarized electrolytic capacitors.

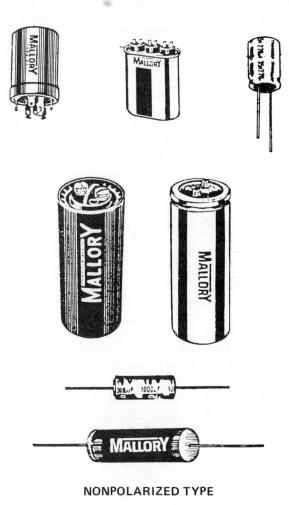

NONPOLARIZED TYPE

FIGURE 2-14 Electrolytic capacitors

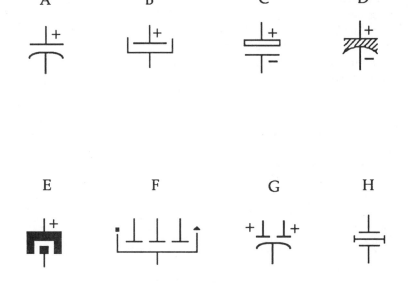

FIGURE 2-15 Electrolytic capacitor symbols

Variable or Adjustable Capacitors

Variable capacitors are needed in electronic circuits to perform selective functions. The most familiar application of the variable capacitor is the tuning capacitor utilized in radio receivers for selecting the stations transmitted on the commercial broadcast band. As the tuning knob is rotated, the capacitor changes in value causing its associated circuits to tune in the signal from the desired station. An example of this type of variable capacitor is illustrated in figure 2-16.

The *two-ganged variable capacitor* consists of two sets of plates that mesh as the shaft is turned. Several flat pieces of metal are connected to form each plate. Air is the dielectric.

The term, ganged, identifies components that are controlled by the main mechanical shaft. That is, if one section is varied by the rotating motion of the main shaft, the other section is also varied. When the shaft is rotated to the point where the movable set of plates (the *rotor*) is entirely meshed with the stationary plates (the *stator*), the capacitance value of the capacitor reaches its maximum. When the shaft is rotated until the rotor extends out of the stator, the value is at its minimum. Miniature versions, often enclosed in plastic, are employed in miniature transistor radios.

Another type of variable capacitor, the *trimmer*, is also shown in figure 2-16. The trimmer's capacitance is varied by turning a screwdriver adjustment. Once the desired value of capacitance is set, it is not normally changed except during alignment procedures.

The graphic symbols used to identify these variable capacitors are illustrated in figure 2-17. Symbol A is the general symbol for any adjustable capacitor. Symbol B represents the two-ganged capacitor used for a radio's tuning selector. Symbol F represents a variable capacitor with a *split stator*; that is, the stator plates are divided into two separate sections, but the rotor plates are not. Symbols C, D, E, and F represent trimmer capacitors. The variations of these trimmer symbols represent specific conditions. For example, symbol D signifies a preset adjustment—once set, the trimmer is not changed except for alignment.

COILS OR INDUCTORS

Inductors are commonly referred to as chokes. As the word implies, inductors choke, or oppose, any change in current. The property of opposing changes in current is known as *inductance*. Whenever current flows through a conductor, magnetic lines of force are produced around the conductor. As long as the current is not a varying (ac) current, the magnetic field is stationary. However, if the current changes in value, the strength of the magnetic field also changes. The change in magnetism exhibited by any current-carrying wire is an example of the property of

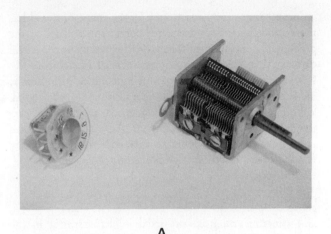

A

B

FIGURE 2-16 Variable capacitors (A) Two-ganged variable (B) Trimmer

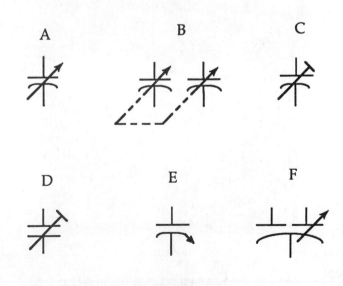

FIGURE 2-17 Variable capacitor symbols

Graphic Symbols of Common Electronic Components 33

inductance. A conductor wound in the shape of a coil has more inductance than a straight conductor because of the greater concentration of the magnetic lines of force. A coil of wire, therefore, is the simplest form of inductor.

The unit of measurement for inductance is the *henry*; however, as with the farad, smaller units are often needed. Common units for specifying inductance are the *millihenry* (*mH*) and the *microhenry* (μH).

Inductors are identified by their core materials, and by whether their values are fixed or adjustable.

Fixed Air-core and Iron-core Inductors

Air-core and iron-core fixed inductors differ in the construction of their cores. The simplest coil is constructed by winding wire in a series of loops around a non-metallic form. As long as the form is not capable of being magnetized, the inductor produced by this method is regarded as the air-core type. Some common air-core forms are plastic, paper, and phenolic.

Two types of air-core fixed inductors, the exposed loop inductor and the molded inductor, are shown in figure 2-18. The molded type may or may not be color coded with the value. In any case, the value is usually given either next to the symbol on the schematic or in the parts list. It is always in millihenry or microhenry units, never in henrys.

The graphic symbols for air-core chokes are illustrated in figure 2-19. The loops may be drawn open as in symbol A or closed as in symbol B. If a coil is shielded within a metallic housing, symbol C or D can be used to identify it.

Iron-core inductors are usually constructed of layers of thin sheets of iron called *laminations*. Insulated copper wire is wound in layers around the core and each layer is separated by additional insulation. This method of winding reduces the core losses inherent in this type of core construction.

Coils of this type are able to provide a large amount of inductance, usually in the henry range. Due to the iron-core material, they are also larger and heavier than air-core inductors.

The symbols for iron-core chokes are illustrated in figure 2-20. Symbols A, B, C, and D are the most common symbols used in modern circuit diagrams. Symbols E and F are used for *tapped chokes* which have taps connected to different points on the windings. The taps permit only certain sections of the total coil to be used.

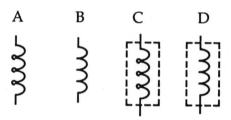

FIGURE 2-19 Symbols for fixed air-core coils

A

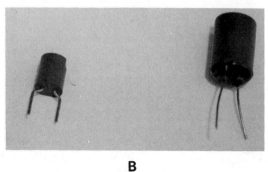

B

FIGURE 2-18 Air-core fixed inductors: (A) open loop, fixed air-core coil, (B) molded, fixed air-core coil

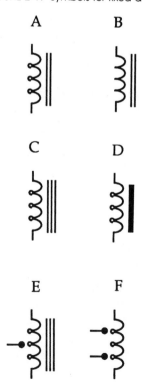

FIGURE 2-20 Symbols for fixed iron-core chokes

34 Electronic Component Symbols

Adjustable Air-core and Iron-core Coils

It is often desirable to adjust the inductance of a coil. A coil can be made adjustable by changing the permeability of the core material, by changing the position of the tap along the total coil value or by expanding or compressing the turns to change the spacing between them. The symbol used on the schematic indicates the type of adjustable coil used in the circuit.

The symbols for adjustable coils are illustrated in figure 2-21. Symbols A and B are often used to depict a coil with a slider that moves along the turns to the required value. This action is very similar to the action of a potentiometer. Symbols C, D, and E, which are interchangeable, represent an adjustable inductor in which the value can be varied by increasing or decreasing the spacing between the turns.

Adjustable iron-core chokes are comparatively rare in modern electronic equipment. Usually, instead of being continuously variable, a series of taps along with a switch are employed to make connection to the specific value. Some types of adjustable coils have cores of ferrite (iron) or powdered iron composition. To change the inductance of this type of adjustable coil, the core is usually moved in or out of the unit. This method of adjusting the inductance is known as *permeability tuning*.

The symbols used to identify permeability-tuned coils are shown in figure 2-22. Symbol A is the most common. The core symbol can be represented by two or three dashed lines. Notice that most of the symbols include an

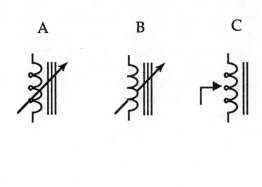

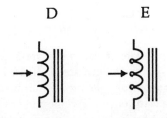

FIGURE 2-21 Symbols for adjustable inductors

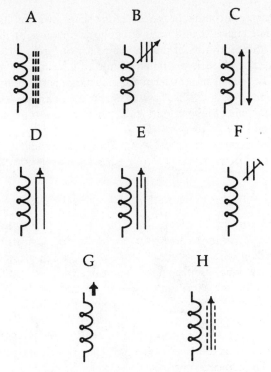

FIGURE 2-22 Symbols for permeability-tuned coils

arrow. The values of most permeability-tuned coils are in the microhenry to millihenry range.

TRANSFORMERS

A *transformer* is two coils of wire wound on a common core material. The winding connected to the input voltage source or circuit is the *primary* winding. Any other sets of windings are the *secondary* windings. As a result of this construction, the magnetic field produced by one set of windings cuts across the other winding or windings. When this occurs, a current is induced in the second winding even though the windings are not electrically connected. This is known as the principle of *mutual inductance*.

The main function of a transformer is to transfer an ac primary voltage to a higher or lower secondary voltage. This action is commonly referred to as *stepping up* or *stepping down* the line voltage. By the principle of mutual induction, a transformer can step up or step down the input voltage by a factor of the *turns ratio* of the input to output winding. That is, if the number of turns on the primary and secondary windings is equal (1:1 turns ratio), the voltage induced in the secondary circuit will be equal to the input voltage. If there are more turns on the secondary than the primary, the induced secondary voltage will be higher. Conversely, if the secondary has fewer turns than the primary, the induced secondary voltage will be lower.

As a result of the step-up or step-down transformer action, the secondary current and voltage will vary exactly in step with the current or voltage in the primary circuit. This transformer phenomenon is utilized for coupling

Graphic Symbols of Common Electronic Components 35

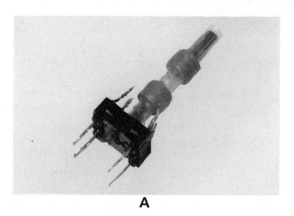

A

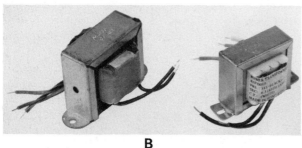

B

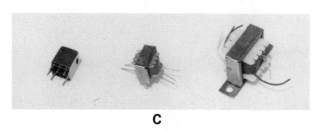

C

FIGURE 2-23 Transformers: (A) RF transformer, air-core type; (B) power transformers, iron-core type; (C) interstage transformer, ferrite-core type

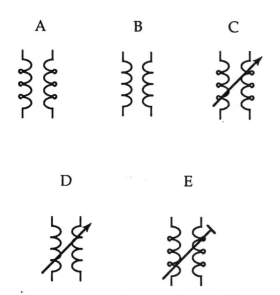

FIGURE 2-24 Symbols for air-core transformers

signals from one stage to the next in receiving or transmitting radio equipment.

Like the inductors (coils) previously discussed, transformers can be constructed with air, ferrite, or iron cores. Different types of transformers are illustrated in figure 2-23.

Air-core Transformers

The symbols for air-core transformers are shown in figure 2-24. Symbols A and B are the basic air-core transformer symbols. Normally, air-core transformers are not adjustable since the only way this can be accomplished is to move the coils closer or further from each other. If these transformers are adjusted by this method, symbols C, D, and E are used.

Iron-core Transformers

Iron-core transformers can be divided into three general categories: power; coupling or interstage; and output. Regardless of the functions of these iron-core transformers, the graphic symbols that represent them are very similar. This makes identifying the specific function difficult when reading a schematic diagram. However, with an understanding of the basic circuit configurations, the trained technician should be able to identify the type of transformer at a specific location in the schematic diagram.

The symbols for iron-core transformers are illustrated in figure 2-25. Symbols A, B, C, D, and E are different ways of representing an iron-core transformer that has one primary and one secondary winding. The primary winding is normally on the left of the symbol, and the secondary winding is on the right. Symbol F represents a multi-secondary transformer that has three separate secondary windings, one of which contains a center tap.

Symbol G represents an *autotransformer*. In an autotransformer, the portion of the transformer from one end to the tap acts as the primary winding, and the complete winding (including the two extreme leads) acts as the secondary winding. Autotransformers can be used to step up or step down voltage by changing the position of the tap relative to the complete winding.

Symbols H and I represent variable iron-core transformers. As with iron-core chokes, these types are seldom encountered; tapped windings are usually employed instead.

Powdered-iron-core Transformers

Powdered-iron-core transformers are available from very small to large units as illustrated in figure 2-26. These transformers are interstage, or tuned, transformers. Their primary function is to couple a specific band of frequencies to which the primary or secondary circuit has been tuned. Tuning these transformers is accomplished by varying the

36 Electronic Component Symbols

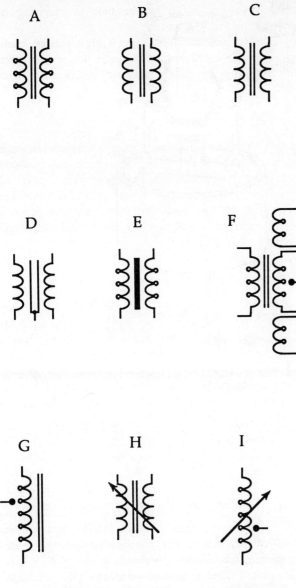

FIGURE 2-25 Symbols for iron-core transformers

permeability of the core. Powdered-iron-core transformers are identified as *i-f* (*intermediate frequency*) transformers. They are used often in rf (radio-frequency) transmitting or receiving equipment.

While most of these transformers are enclosed in a metal can for shielding, some may not have this covering. Fixed capacitors connected across the primary and/or the secondary windings may also be enclosed in the same housing.

The symbols used to designate powdered-iron-core transformers are illustrated in figure 2-27. Symbols A, B, and C represent nonadjustable windings. Symbols D through L represent transformers in which each winding is individually adjustable. Symbols M through R represent transformers in which one adjustment affects both windings.

The reference designation L is used for all inductors (coils), while the letter T identifies transformers.

FIGURE 2-26 Powdered iron-core transformers

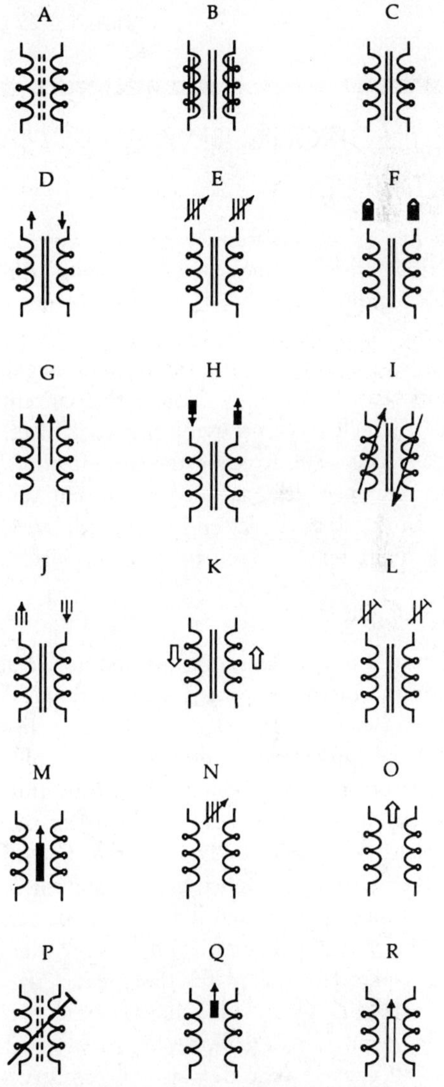

FIGURE 2-27 Symbols for powdered iron-core transformers

Graphic Symbols of Common Electronic Components

 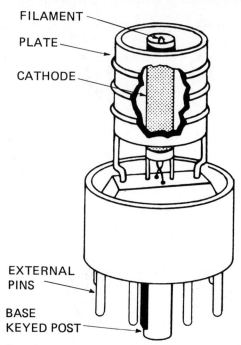

FIGURE 2-28 Construction of a vacuum tube diode

UNIT 2 ELECTRONIC DEVICES
VACUUM TUBES

Vacuum (electron) tubes were developed to control current flow which is measured in *amperes*. *Electrons* are negatively charged particles that surround the positively charged nucleus in an atom. If a complete conductive path is available, electrons will flow from a negative to a positive potential (voltage). Unless this current flow is controlled, it will serve no useful purpose. Electron tubes provide a means of controlling the current flow.

The reference designation for vacuum tubes is the letter V. In the past, the letter T was used, but in modern schematic diagrams T is used for transformers.

Diode Tubes

The vacuum tube diode was the first device developed to provide the means to control current flow. The diode consists of two primary elements—the plate and the filament. When an external voltage source is applied to the filament, electrons are "boiled" off from this element which is coated with a material capable of releasing many free electrons when heated. If a positive potential (compared to the filament) is applied to the plate, the "cloud" of electrons formed around the filament will be attracted to the positively charged plate. The larger the positive potential applied to the plate, the greater the electron movement to the plate. This fundamental principle of electron movement and control by an external voltage applies to all vacuum tube devices. The construction of a diode tube is illustrated in figure 2-28.

The vacuum tube diode is used primarily for an important electronic concept known as *rectification*. Rectification is the conversion of alternating current to pulsating direct current. The pulsating dc voltage is then applied to a capacitor to complete the smoothing action known as filtering. This function is basic to electronic power supply circuits.

Schematic Symbols. The schematic symbols that represent vacuum tube diodes are shown in figure 2-29. The circle around symbol A represents the glass or metal envelope. The pointed portion at the bottom of the circle represents the filament and the straight horizontal bar with the vertical line extending upward represents the plate.

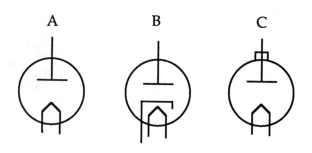

FIGURE 2-29 Symbols for vacuum tube diodes

38   Electronic Component Symbols

In symbol B, an additional element called the *cathode* is placed between the filament and plate. The cathode is a metal sleeve that is coated with a special electron-emitting material. It fits over the filament, changing the function of the filament. Now the filament is used only to heat the cathode. In this type of tube, the filament is more properly called the *heater*. When the cathode element is added, a more abundant source of electrons is produced. In addition, the emission of electrons is not affected by an ac voltage source, which is often used as the primary heating source.

In the diode represented by symbol C, the plate extends beyond the top of the tube to a metal cap instead of to pins at the base.

Tube Basing Diagrams. Internal connections from the tube elements within the enclosed glass or metal envelope are made to pins in its base. The common bases for modern tubes are the seven-pin and nine-pin miniatures, the eight-pin octal, and the twelve-pin compactron. In the seven-pin, nine-pin, and twelve-pin types, the pins are symmetrically spaced and have a blank space which serves as a guide for inserting the tube into a convenient socket. In some tubes, not all pins are connected to elements within the enclosed envelope. In fact, some pins may be omitted from the base if a locating key is employed at the base of the tube.

Each pin is numbered for reference purposes. With the base of the tube pointed towards you, pin 1 is the first pin starting clockwise from the blank space or the locating key (see figure 2-30). All other pins are numbered consecutively, reading clockwise around the base. If a pin is omitted on the tube base, the number ordinarily assigned to the missing pin is skipped, but the numbered locations of the other pins remain the same.

These pin numbers are also placed on schematics next to the symbols, at the point where the respective tube elements enter the tube envelope. When the pin numbers are used on a schematic, the schematic is known as a *basing diagram*. By referring to the numbers beside the tube symbol, the service technician is able to locate the pin connected to each element.

Triode Tubes

Triode tubes were developed for the purpose of amplifying small signals, a function which diode tubes are not capable of performing. The triode is made by adding a third element, the *control grid*, to the diode. The addition of the control grid has made amplification of signals possible.

The symbols for triodes are shown in figure 2-31. The grid is represented by the dashed line between the cathode and plate in symbol A. Notice the numbers placed near the points where the symbols for the individual elements enter the envelope. These numbers denote the pin number of the base to which this element is connected.

In the past, the zigzag line, as shown in symbol B, was used to represent the control grid. The principal differences between modern tube symbols and old tube symbols are the heaviness of the lines and the shapes of the figures used to depict the tube elements inside the enclosed envelope. In symbols C and D, the elements of the tube are drawn differently, but they are easily recognized. The elements represented by the symbols may also be tilted, drawn backwards, upside down, or in various other arrangements. The glass or metallic envelope may be represented by a circle as in symbols A, B, and C, or elongated as in D and E. It may even be eliminated entirely. Regardless of the variations of the triode symbol, the basic symbol remains the same.

Tetrode and Pentode Tubes

Deficiencies in the triode led to the development of the tetrode tube; deficiencies discovered subsequently in the tetrode led to the development of the pentode tube. Due to *interelectrode capacitance* (capacitance between one electrode and the next electrode toward the anode), use of the triode is limited to low-frequency applications. To

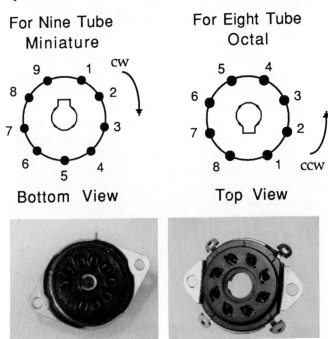

FIGURE 2-30 Pin numbering system

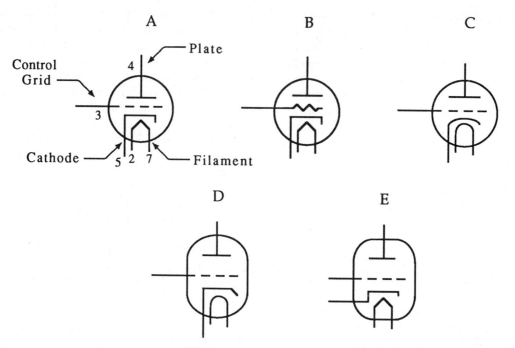

FIGURE 2-31 Symbols for triode tubes

solve this problem, a third grid called the *screen grid* is inserted between the plate and the control grid. The new tube is known as the tetrode.

The shielding action of this additional grid greatly reduces the effect of the electrode capacitance between the control grid and the plate. In addition, tetrode tubes provide a higher gain (amplification) than can be obtained by a triode. The schematic symbol for the tetrode is shown in figure 2-32.

The operation of the tetrode seemed ideal until the phenomenon known as *secondary emission* was discovered. Secondary emission causes electrons collected on the surface of the plate to be bounced off and knocked loose, resulting in a loss of plate current. This phenomenon reduces the effectiveness of the tetrode tube.

The pentode tube solves the problem of secondary emission. It contains a third grid, known as the *suppressor grid*, between the plate and the screen grid. This additional grid greatly improves the performance of the vacuum tube as an amplifier.

Although its operation is more like that of a tetrode, the *beam-power tube* is often identified as a pentode. Instead of a suppressor grid, however, it has two metal vanes known as the *beam-forming plates*, which are positioned to aid, or focus, the electrons toward the plate. These plates are electrically connected to the cathode. As a result of this specialized tube element, much greater power can be delivered to the load.

The symbols for the pentode and the beam-power tube are shown in figure 2-33. Pentode symbol A is most often used for the beam-power tube, but sometimes the tetrode symbol or symbol B is used. Regardless of which symbol is used, a tube manual should be consulted to determine whether the tube in question is a beam-power tube or a pentode.

Pentagrid Tubes

Pentagrid tubes contain five grids between the cathode and the plate. The most common application of

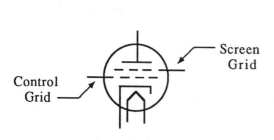

FIGURE 2-32 Schematic symbol for a tetrode

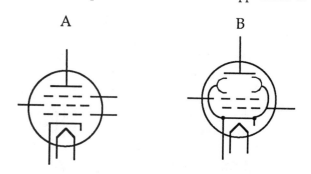

FIGURE 2-33 Pentode and beam-power tube symbols

40 Electronic Component Symbols

pentagrid tubes is in a superheterodyne radio receiver. The first grid, which is closest to the cathode, acts as an oscillator grid and the second grid acts as the plate element for the oscillator section. Electrons flow towards the plate in a conventional manner.

The third grid, which is the control grid, functions as a conventional pentode. This grid receives the signal from the antenna circuit. The stream of electrons from the oscillator grids varies in step with the oscillator frequency. The signals from the oscillator and the antenna beat, or *heterodyne*, together to produce a new signal which varies in amplitude with the antenna signals, but has a frequency equal to the difference between the oscillator and the antenna signals.

The fourth and fifth grids, which are the suppressor and screen grids, function as in a conventional pentode.

Other multifunction tubes often found in tube-operated equipment are the duodiode (two diodes in one tube housing), the duotriode, the duodiode with a triode or pentode, and the triode-pentode combination. In the schematic symbols that represent these multifunction tubes, the tube elements are often drawn side by side. The symbols for pentagrid and multifunction tubes are illustrated in figure 2-34.

The dashed lines through the center of symbol B represent a shield placed between the two sections for isolation. Although some tubes may not have this shield, they are separated on the schematic diagram to show their functional positions.

In symbol C, the two halves of a duotriode are separated, but they are enclosed in the same tube envelope. This signifies that the duotriode is two triodes housed in one envelope. The dotted lines are omitted and the sides of the envelope are left open in symbol D. This symbol can also be used to represent multifunction tubes in different locations on the schematic diagram.

Symbol E represents another multifunction tube, the dual-pentode. In this tube, the cathode, control grid, and screen grid are common to both sections. In some tubes, though, only the cathode may be shared. If the cathode is the only shared element and if each section is drawn at a different location on the schematic, the cathode is shown in both places, but the same pin number is placed beside each cathode symbol on the diagram.

Gas Tubes and Phototubes

Instead of being evacuated, certain tubes, known as gas tubes, are filled with an inert gas. Tubes of this type require no filament. They are sometimes called *cold-cathode* or *ionically-heated cathode* tubes. Other tubes, known as phototubes, contain a photocathode. The symbols for gas tubes and phototubes are shown in figure 2-35.

Notice that a circle, rather than a flat bar, is used to represent the cathode. No filament is represented. Sometimes the circle is replaced by a dot as in symbol B. In either case, a smaller dot, indicating a gas-filled tube, also appears inside the envelope. The other connection shown inside symbol B is a wire between two pins on the tube base. This wire acts as a switch to open another circuit if this tube is accidentally removed. This protective feature is normally included in high-voltage applications.

Symbol C represents a gas-filled tube known as a *thyratron*. Notice that it is similar to the triode except that it has a dot inside the envelope. The thyratron is usually used in electronic control circuits.

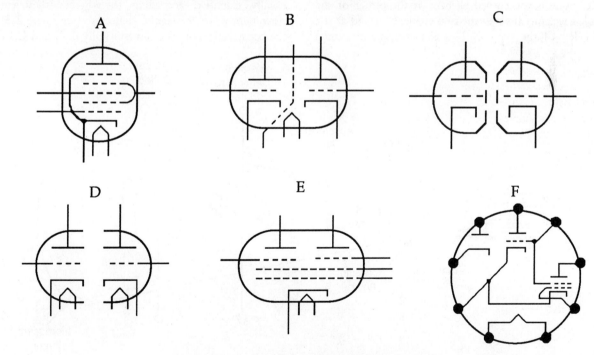

FIGURE 2-34 Pentagrid and multifunction tube symbols

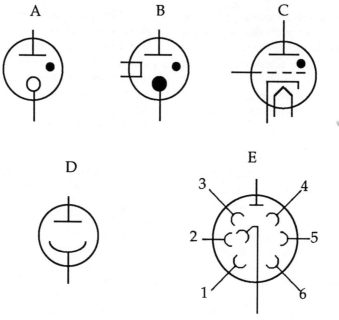

FIGURE 2-35 Gas tube and phototube symbols

Symbol D depicts a phototube. Materials such as sodium, cesium, and potassium emit electrons when exposed to light. A curved piece of metal coated with a photosensitive material acts as the cathode. The plate, which is a metal rod situated behind the cathode, acts as the collector. When light strikes the photosensitive cathode, electrons are released and flow to the plate. The curved portion of the symbol represents the cathode and the flat bar represents the plate, or anode.

Symbol E represents a *photomultiplier* which functions the same as the phototube represented by symbol D. The main cathode is the curved element at the center of the symbol. Electrons flow to the first element located at the bottom left of the symbol. This element, in turn, emits more electrons toward the adjacent element. Each of these elements attracts the electrons and each subsequently emits more electrons than the preceding one. The conventional anode symbol is used for the element from which the output is taken.

Cathode-Ray Tubes (CRTs)

Cathode-ray tubes are used as oscilloscope displays and, to a larger extent, as television picture tubes. They are usually classified according to whether they employ *electrostatic* or *electromagnetic* deflection (see figure 2-36 for the construction of an electrostatically-deflected CRT).

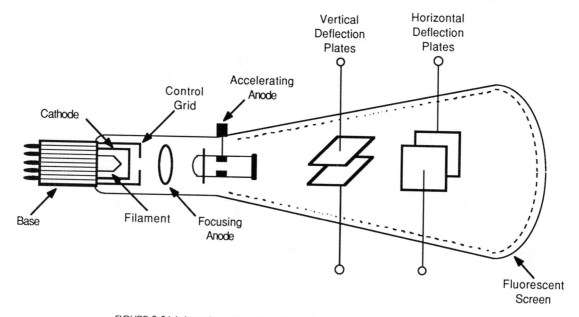

FIGURE 2-36 Internal construction of an electrostatically-deflected CRT

42   Electronic Component Symbols

The operating principle of the electrostatically-controlled CRT is similar to that of the conventional vacuum tube devices discussed previously. The electrons emitted from the cathode are attracted to the positive voltage on the accelerated anode. First, however, they must flow through the control grid, which is in the form of a metal cylinder with a hole in one end. A negative voltage applied to the control grid controls the amount of electrons passing through the grid. From the grid, the electrons enter the *focusing anode*, which concentrates the electron stream into a narrow beam. The velocity of this beam is then increased by two high-voltage, accelerating anodes. This concentrated electron beam next encounters the pair of deflection plates. One pair of plates moves the beam in a vertical (up-down) direction while the second pair deflects the beam in a horizontal (left-right) direction.

After the beam passes through the deflection plates, it strikes the phosphorous-coated screen with great force causing the screen to display an image on the face of the tube. The brightness of the image depends on the number of electrons striking the screen and the velocity of the electron beam.

Oscilloscopes and other electronic test equipment which require the display of signal amplitudes and waveshapes employ this type of CRT.

The schematic symbols sometimes used to depict electrostatic, cathode-ray tubes are shown in figure 2-37. In symbol A, the tube elements are shown in the order in which they appear in the tube. From the left side of the symbol, the heater, the cathode, the control grid, the focusing anode, and the accelerating anodes are represented. The two sets of deflection plates are depicted by the Y-shaped symbols that follow the last accelerating anode.

Symbol B is also used on schematic diagrams to identify an electrostatic CRT. Notice that in this symbol the focusing anode is placed between the two sections of the accelerating anode. When other arrangements are used, the schematic symbol usually shows the actual arrangements of the tube elements, or electrodes.

Electromagnetically-deflected CRTs are commonly used as TV picture tubes. They are similar to electrostatic CRTs, but they lack deflection plates. To deflect the electron beam horizontally or vertically, they have a set of coils located around the neck of the tube.

The symbols for electromagnetically-deflected CRTs are shown in figure 2-38. Symbol A represents an electromagnetically-deflected CRT that uses electrostatic focus. The connection to the line inside the envelope in the lower right-hand portion of the symbol is the second

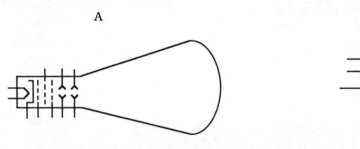

FIGURE 2-37 Electrostatic CRT tube symbols

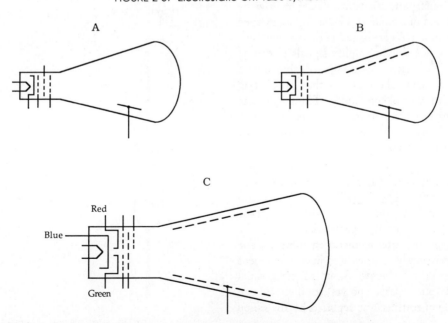

FIGURE 2-38 Electromagnetically-deflected CRT tube symbols

Graphic Symbols of Common Electronic Components 43

anode, or high-voltage connection. This connection is made to a contact on the side of the tube instead of through the base to isolate it from the pins on the tube base. The focusing electrode is omitted in symbol B implying that the focusing is accomplished externally.

Symbol C represents a color-TV picture tube. This symbol is similar to those used for black-and-white picture tubes, except that it has three separate electron "guns." The elements from left to right are the heater, the cathode, the control grids, the screen grids, and the focusing electrode. Notice that the focusing electrode is common to all three guns. The high-voltage connection is at the bottom right.

SEMICONDUCTOR DEVICES

Semiconductor devices are characterized by conductivity ranging between that of an insulator and that of a conductor; at low temperatures, conductivity is almost absent and at high temperatures, it is almost metallic. The semiconductor materials used most often in semiconductor devices are silicon and germanium. Like vacuum tube devices, semiconductor devices are used to control current. Semiconductor devices are also known as solid-state devices.

Semiconductor Diodes

The simplest semiconductor device is the pn-junction diode which was developed in the infant stages of this rapidly moving technology. The semiconductor diode, like its vacuum tube counterpart, can control current flow by allowing conduction in one mode and behaving as an open switch in the other mode. As explained in the vacuum tube section, the diode is not able to amplify signals due to its electrical action. The major advantage of this device compared to the vacuum tube diode is that no heating source is necessary to control conduction. The minute size and "solid" properties of this device are other attractive features which have virtually eliminated the use of the vacuum tube diode for rectification.

Power Rectifiers. The power supply rectifier is a type of pn-diode. In the power supplies of radios, televisions, and other electronic equipment, the ac line voltage must be converted to dc voltage to operate the various stages of the electronic instrument or entertainment equipment. The oldest type of power supply rectifier used in entertainment equipment was made of *selenium*. Some of these units actually contained four separate power rectifiers in one housing. The selenium rectifier was replaced by the *silicon* rectifier.

Schematic symbols for silicon rectifiers are illustrated in figure 2-39. Symbol A is the universal symbol for power rectifiers. In some cases, such as in symbols B and C, part of the symbol is not shaded. In other cases, such as in symbol D, the entire symbol may be left unshaded.

There is much disagreement among manufacturers on which reference designation to use for power rectifiers on schematic diagrams. The letters *D* and *CR* are the most common; however, the letters *SE*, *X*, *E*, and *REC* are also encountered.

Signal Diodes. Signal diodes, like power rectifiers, are used to rectify an alternating voltage to a direct voltage. The major difference between signal diodes and power rectifiers is in the electrical function each performs in a circuit. Power rectifiers are used in power supply circuits to handle large amounts of current and, at times, high voltages. Signal diodes are normally used in circuits where low levels of current and voltage are present. Typical signal diodes are shown in figure 2-40. The schematic symbols for signal diodes are identical to those used for power diodes.

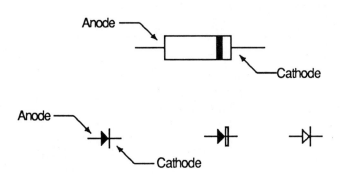

FIGURE 2-39 Silicon rectifiers and schematic symbols

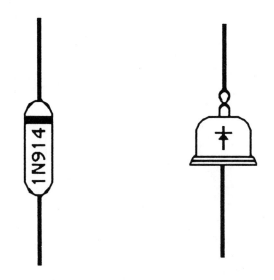

FIGURE 2-40 Typical small signal diodes

Zener Diodes. Zener diodes are designed to operate under conditions in which other diodes, such as the power diode, might be damaged. When a reverse voltage is applied to diode elements, current flows through the diode in the reverse-biased mode, which can damage the diode. When the zener diode is reverse-biased, however, it is able to control the reverse current. Zener diode case styles are very much like power diode and signal diode case styles. Its characteristics are very similar to those of the power or signal diodes.

In practical circuit applications, zener diodes serve as *voltage regulators*. That is, when the applied voltage exceeds a given value, the zener diode conducts, limiting the voltage to the desired value. The symbols used to denote a zener diode are shown in figure 2-41. Symbols A and C are probably the most common. The basic symbols, such as A, B, E, F, and G, may have a circle.

Transistors

The invention of the transistor by Drs. William Shockley, John Bordeen, and Walter H. Brattain of Bell Telephone Laboratories (1948) revolutionized the world of electronics. The word transistor is a broad term applied to the entire field of semiconductor devices that have three or more leads or terminals. Transistors come in many sizes and shapes, but each transistor has its own particular application. Some of the applications are amplification, oscillation, switching, and mixing.

The Bipolar-Junction Transistor (BJT). The bipolar-junction transistor corresponds roughly to the triode vacuum tube. The basic bipolar transistor elements are illustrated in figure 2-42. Notice that both types of transistors contain two pn junctions as in a simple diode. Just as adding the grid to the vacuum tube makes it possible to control a large voltage with a small voltage applied to the grid, adding a third layer to the basic semiconductor diode makes it possible to control a large current with a much smaller input current applied to this additional element. This element is known as the *base*. Transistors are current-controlled devices, while vacuum tubes are voltage-operated devices. The other two elements of a basic bipolar transistor are the *emitter* and the *collector*.

The symbols for npn and pnp transistors are shown in figure 2-43. The only difference between the symbols for npn transistors and those for pnp transistors is the direction of the arrowhead, which represents the emitter element.

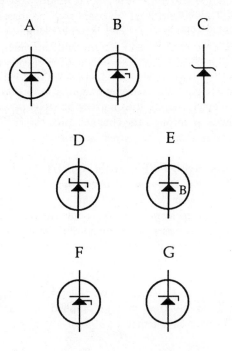

FIGURE 2-41 Zener diode symbols

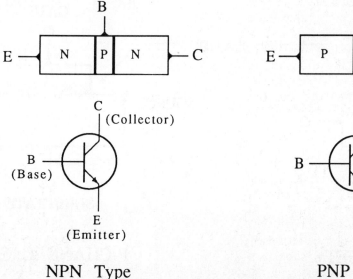

NPN Type

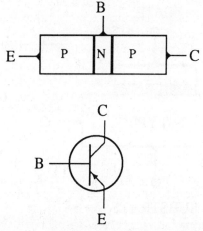

PNP Type

FIGURE 2-42 Basic pnp and npn transistors

NPN Transistors

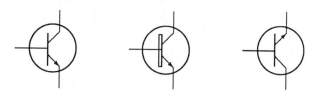

PNP Transistors

FIGURE 2-43 Symbols for npn and pnp transistors

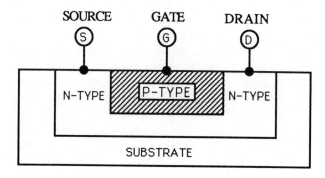

FIGURE 2-44 Construction of a JFET

Field-effect Transistors. Field-effect transistors (FET) are more like conventional vacuum tubes than transistors in that they are voltage-operated devices. There are two general types of FETs: the *junction FET* (*JFET*) and the *insulated-gate FET* (*IGFET*). The construction of the JFET is shown in figure 2-44.

The substrate, or *channel*, material in this type of JFET is an n-material. A p-material is then diffused into the n channel material forming a pn junction between the channel areas. The connection to one end of the bar is called the *source* while the opposite connection is the *drain*. The third element of this device is the *gate*, which consists of the p-material for this n-channel JFET. The channel can be made of either an n-type or a p-type material as illustrated in figure 2-44.

The construction of the IGFET is illustrated in figure 2-45. Here, the junction gate is replaced by a small metal plate which is electrodeposited on top of an insulating film on the face of the substrate bar. When a varying voltage is applied to the gate, the resulting magnetic field penetrates into the channel, causing it to narrow or widen to control the current flow through the channel. Like the JFET, the channel material may be either an n-type or a p-type material.

Another type of FET in the IGFET family is the *MOSFET* (*metal-oxide-semiconductor*) which has become increasingly popular in many electronic circuit applications. Its name is derived from the unique construction of its gate region, which consists of a metal-film gate, an oxide insulation, and a semiconductor wafer. In an n-channel MOSFET, the channel is n-type material and the source and drain sections are two diffused p-type regions. Current from the source to the drain is controlled by the magnetic field produced by the voltage applied to the gate. MOSFETs may have either a single or a dual gate.

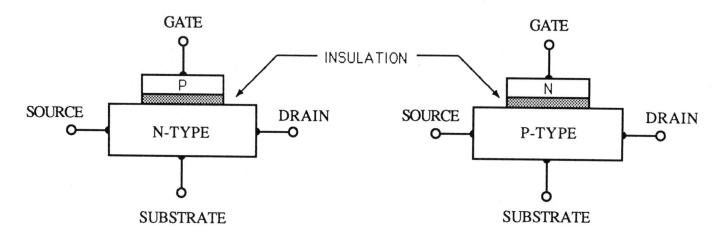

N-CHANNEL IGFET   P-CHANNEL IGFET

FIGURE 2-45 Construction of an IGFET

FET symbols are illustrated in figure 2-46. Symbols A and B are JFETs; symbol A is the n-channel JFET and symbol B represents the p-channel JFET. Symbols C and D represent the n-channel and p-channel IGFETs respectively. Symbol E is for an IGFET with dual gates.

The variations in the IGFET devices result from the substrate material used. Connections to the substrate material may be internal, external, or omitted as in symbols C and D. A connection to the substrate material is represented by the arrowhead within the symbol. The arrowhead would be reversed for the p-type IGFET. Symbols J and K are other variations of the JFET symbol.

Unijunction Transistors. The unijunction transistor (UJT) is constructed much like the FET. The substrate material can be either a p-type or an n-type material. Connections are made to each end of the bar for the *two* bases. A single pn junction is formed approximately one-quarter of the distance from one end of the bar. Connections are made to the two bases and to the spot of p-material that serves as the emitter.

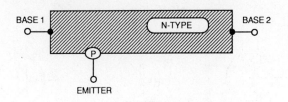

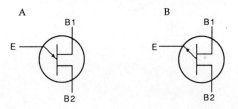

FIGURE 2-47 UJT construction and schematic symbols

In operation, the UJT does not conduct until a certain voltage level is applied across the base-emitter junctions. When the voltage level is applied, it suddenly turns on. The basic construction of the UJT and the symbol for the UJT are illustrated in figure 2-47. Symbol A represents a UJT with an n-type base material and symbol B represents a UJT with a p-type base.

The reference designation assigned to transistors is the letter *Q*. Nevertheless, the letters *X, V, T,* and *TR* are also employed on many schematic diagrams.

Special Semiconductor Devices

A group of very special devices was developed to perform specific functions which the basic transistor devices could not perform satisfactorily. These specialized devices include *thyristors, triacs, tunnel diodes, varactor diodes, photodiodes,* and *diacs*.

Thyristors. Thyristors are semiconductor devices that act as switches. Their *bistable* action (switching action of on or off) depends on pnpn regenerative feedback. Thyristors can be two-terminal, three-terminal, or four-terminal devices. They are available in both unidirectional and bidirectional devices. The silicon controlled rectifier (SCR) is the best known of all thyristor devices. It is used predominantly to control ac line voltage. For example, the SCR is used as the electronic device that controls light dimmers and variable speed, line voltage driven tools.

Symbols for thyristors are shown in figure 2-48. SCRs are represented by symbols A and B. These devices are unidirectional in that they will not conduct until a trigger current is applied to the gate (G) terminal. Symbol C represents a gate turn-off SCR. The gate current for this type of SCR will turn the unit off instead of on. Symbols A, B, and C are for units with n-type gate material. Symbols D, E, and F are for the same type of units, except they contain a p-type gate material. Symbols G and H are for a reverse-blocking, tetrode-type, semiconductor-controlled switch.

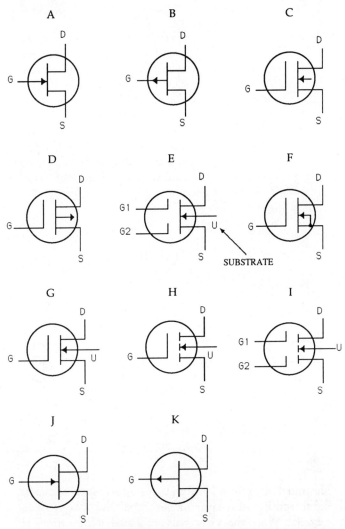

FIGURE 2-46 FET symbols

Graphic Symbols of Common Electronic Components 47

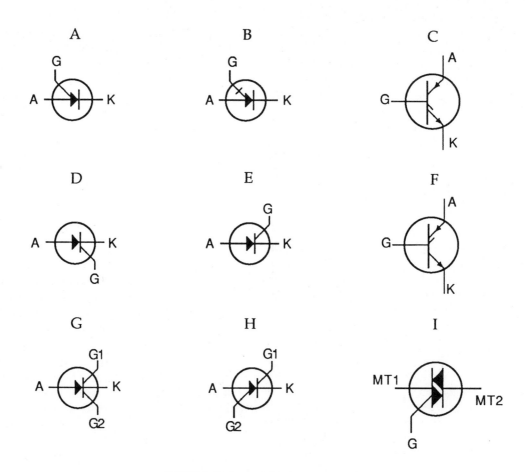

FIGURE 2-48 Symbols for thyristors

Symbol I represents a bidirectional thyristor (three-terminal type), which is also called a *triac* or gated switch. The triac conducts in both directions in response to a positive or negative gate signal. The SCR, on the other hand, conducts to the polarity of the gate material—that is, the positive signal to the p-type gate and the negative signal to the n-type gate.

Tunnel Diodes. The electrical characteristics of tunnel diodes are such that they reverse their conduction with small applied voltages, but return to their normal conduction cycle with any increase in the applied voltage. This unusual characteristic, combined with their high switching speeds, makes tunnel diodes useful devices in switching, amplifying, and oscillating circuits, even at microwave frequencies. These diodes consume very little power and are relatively unaffected by severe environmental temperature changes. The symbols used to depict tunnel diodes are shown in figure 2-49. As with the other semiconductor devices, the circle may or may not be used.

Varactor Diodes. Varactor diodes are also known as capacitive diodes, reactance diodes, parametric diodes, or varicaps. The schematic symbols for these devices are illustrated in figure 2-50. Recall that a basic semiconductor diode consists of a layer of p-material and a layer of n-material. When these materials are joined, a layer forms at the junction, which serves as an insulator. This condition is

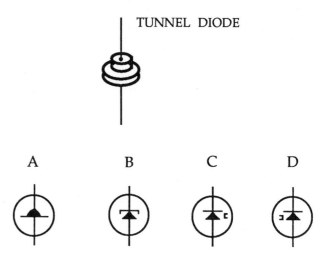

FIGURE 2-49 Tunnel diode and symbols

48 Electronic Component Symbols

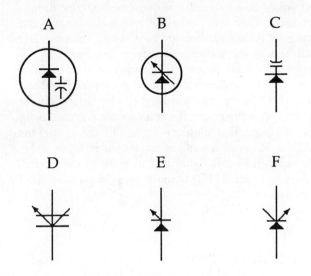

FIGURE 2-50 Symbols for varactor diodes

the same as for a capacitor, which consists of two conductive plates separated by an insulating material. In the normal diode, steps are taken to minimize this inherent capacitance. In the capacitive diode, however, the capacitance is emphasized. The capacitance of these diodes varies with the voltage across the diode. Thus, by changing the voltage level, the capacitance value can be varied and a circuit can be tuned with this device.

Photodiodes. All semiconductor materials are sensitive to light due to their construction and material composition. In certain applications, they may function as resistors in which their values vary according to the amount of light striking their surfaces. In this use, they are known as *light-dependent resistors* (*LDRs*) or photodiodes. Symbols for LDRs are illustrated in figure 2-51.

In addition, a voltage can be generated when light strikes the diode. In this application, it is known as a *solar cell*. In symbols A, B, and C the arrows, the letter L, and the Greek letter lambda (λ) signify that the device is light sensitive. Symbol D represents the solar cell. The arrows are added to the basic symbol for a cell to signify the cell's sensitivity to light.

Some diodes emit light when voltage is applied to their terminals. These diodes are known as *light-emitting diodes* (*LEDs*). Symbol E is for a photoemissive LED. Note that it is the same symbol as A except for the direction of the arrows.

Integrated Circuits

All of the semiconductor devices discussed to this point are *discrete components*—that is, each device is a separate, distinct item. All discrete semiconductor devices are fabricated in large numbers on tiny semiconductor substrate chips. Several hundred transistors, diodes, resistors, and other active or passive components can be impressed on a single chip. As a result of these high-density fabrication methods, the development of *integrated circuit* (*IC*) technology has revolutionized modern electronic circuitry. The ability in the latest integrated circuit manufacturing technology to increase the successful yield of LSI (large-scale integration) has virtually eliminated large complicated circuitry using discrete components such as transistors, diodes, resistors, and capacitors.

The basic integrated circuit device is normally represented on schematic diagrams by the triangular, square, or rectangular outline symbol as shown in figure 2-52. The internal circuitry of the IC is usually not shown because of the complexity of the circuits within the miniature package. Also, some of the circuitry is considerably different than

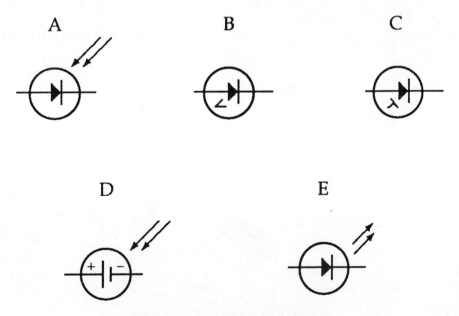

FIGURE 2-51 Symbols for photodiodes

Graphic Symbols of Common Electronic Components 49

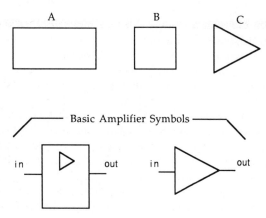

FIGURE 2-52 Integrated circuit symbol outlines

what would be employed if conventional discrete devices were used. Therefore, only the triangular, square, and rectangular outline symbols are used, with leads extending from them to show connections from any external components to the IC proper. If the internal circuitry is of interest, a manufacturer's data sheet or a separate drawing that shows the internal circuitry of the device can be provided to complement the main circuit diagram. Schematic diagrams that illustrate the use of the square and triangular IC symbol outlines are shown in figure 2-53. Also included in this figure is a diagram of the internal circuitry of a specific IC obtained from the manufacturer's data sheet.

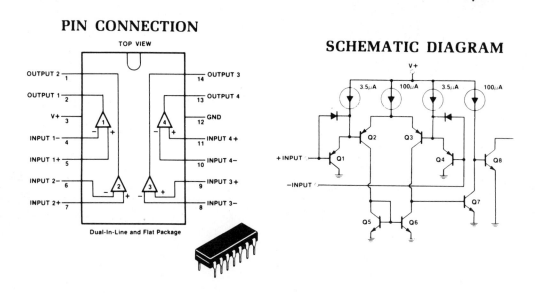

FIGURE 2-53 Integrated circuit diagrams

50   Electronic Component Symbols

UNIT 3 CONTACTS, CONTACTORS, SWITCHES, AND RELAYS

Contacts, contactors, switches, and relays are used to open and close circuits. A circuit is closed when there is a complete path through it enabling current to flow. The process of initiating a simple or very complex electronic function normally requires that a switch be thrown to turn on the equipment. A circuit can also be opened by the simple switching function.

Some means must be provided to close or open many of the circuits in complex electronic equipment. The following devices perform this switching function by a simple mechanical action or with semiconductor devices.

MECHANICAL SWITCHES

The simplest mechanical switch is the *knife switch* (see figure 2-54A). When the arm on the knife switch moves down, it engages the clips at the end and closes, or completes, the circuit to which it is connected.

Another mechanical switch is the *toggle switch* (see figure 2-54B). The action of this switch is similar to that of the knife switch in that switching to the open or closed position is accomplished by a simple lever action. This style of switch is commonly used in residential and commercial buildings to control room lighting. It is also used in electronic equipment.

In the *slide switch* (figure 2-54 C), which is also used in electronic equipment, switching is accomplished by a simple sliding motion which completes the contact closure within the switch housing. This type of switch is available in standard to subminiature sizes.

The *rocker switch* (figure 2-54D) has found increasing use in modern electronic equipment due to its stylish and low-profile appearance. As with the preceding switches, the contact closure is located inside the switch housing. The action of the contact closure in the rocker switch can be compared to the action of the knife switch.

The push-button switch and the rotary switch are also used often in electronic equipment. Due to their selective applications in many control circuits, they will be identified and discussed in detail later in this text.

SWITCH CONTACT IDENTIFICATION

Mechanically-activated switches are identified by the number of contacts (poles) they possess and by the number of positions (throws) they are able to complete. For

A

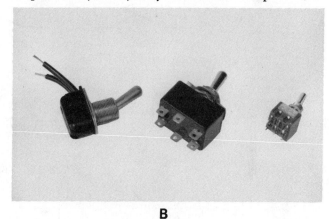
B

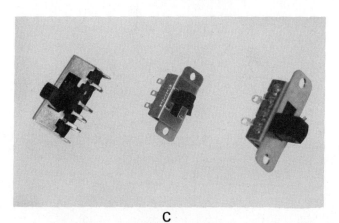
C

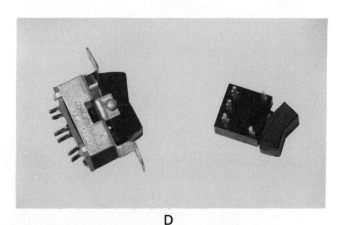
D

FIGURE 2-54 Mechanical switches: (A) knife switches, (B) toggle switches, (C) slide switches, (D) rocker switches

Graphic Symbols of Common Electronic Components 51

example, the *single-pole, single-throw* (SPST) contact is the simplest type of switch available. This switch has two terminals. One terminal is the movable arm, or wiper, which is connected to the movable contact inside the switch housing. The other terminal is the stationary contact terminal which the arm contact meets to complete the contact closing action. Other contacts are identified as single-pole, double-throw (SPDT), double-pole, single-throw (DPST), and double-pole, double-throw (DPDT).

The Single-Pole, Single-Throw (SPST) Contact

The single-pole, single-throw (SPST) contact is the simplest switch contact. It can make connection for only one line, at only one point. Symbols for the SPST switch are illustrated in figure 2-55.

Symbols A, B, and C are similar in appearance. Symbol D is not commonly found in modern schematic diagrams, but it represents the same electrical function as all SPST switch contacts. Symbol E resembles the slide switch style. The arrowhead in the symbol represents the movable arm, or wiper, contact of the actual switch. The circles in the symbols (if present) represent the contact points of the switch.

The Single-Pole, Double-Throw (SPDT) Contact

If an additional contact position is added to the single-pole, single-throw (SPST) contact, the electrical action is such that two separate portions of the same circuits may be controlled by a single switch. This type of switch is identified as the SPDT or *single-pole, double-throw* contact. Schematic symbols for SPDTs are illustrated in figure 2-56.

The Double-Pole, Single-Throw (DPST) Contact

If two separate and independent circuits must be controlled with a single switch, double-pole switches are required. On the other hand, if two separate portions of the same circuit must be controlled with a single switch, a SPDT switch can be used. The *double-pole, single-throw* (*DPST*) is the simplest double-throw switch. It is con-

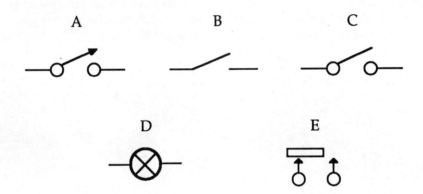

FIGURE 2-55 SPST contact symbols

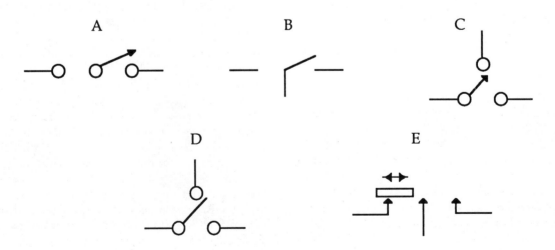

FIGURE 2-56 SPDT contact symbols

52   Electronic Component Symbols

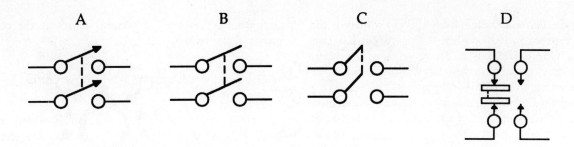

FIGURE 2-57 DPST contact symbols

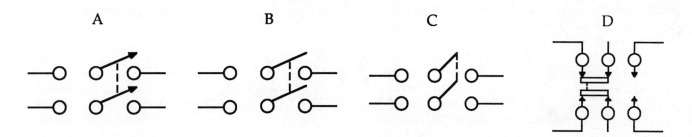

FIGURE 2-58 DPDT switch symbols

structed of two separate SPST switches controlled by the same arm or wiper. The two arms are connected, or *ganged*, together by an insulating material so that when one is moved, the other arm automatically follows. The two arms are mechanically connected but electrically isolated from each other. The symbols that identify DPST switches are shown in figure 2-57.

The Double-Pole, Double-Throw (DPDT) Contact

When another set of terminals is added to the DPST switch so that the two mechanically-connected arms make contact in either of two positions, the result is a *double-pole, double-throw* switch. Symbols for the DPDT switch are shown in figure 2-58.

WAFER SWITCH

The *wafer switch* (see figure 2-59) is used to complete or make connections to more than one point in a circuit controlled from a single source. One contact around the edge is longer than the other contact points so that it always makes connection with the circular ring in the center. Notice that one point on the ring extends out farther than the rest. As the switch shaft is rotated, this point is connected to each contact, one after another.

Wafer switches are also classified as SPST, SPDT, DPST, and DPDP. However, these classifications may not fully identify the variety of combinations possible with

FIGURE 2-59 Wafer switch with single wafer section

this type of switch. The standard system used to identify wafer switches indicates the number of poles (single, double, triple, etc.) followed by the number of contact points or throws—for example, single-pole, six-throw (see figure 2-60A and B).

A single-pole, eight-position (throw) switch is shown at figure 2-60C. The arrows extending from the small circles represent the contacts around the outside of the wafer. The longest arrow is the longest contact. The ring is deliberately drawn so that it does not touch the arrowheads except at the long contact and at the point of extension on the ring. As the switch is rotated, this extension contacts each arrowhead in turn. It is important to note that the symbol is usually illustrated as being viewed *from the shaft end*, and the terminals or contacts are numbered *clockwise*. If the wafer is illustrated from the rear, the contacts will, of course, be numbered counterclockwise.

Graphic Symbols of Common Electronic Components 53

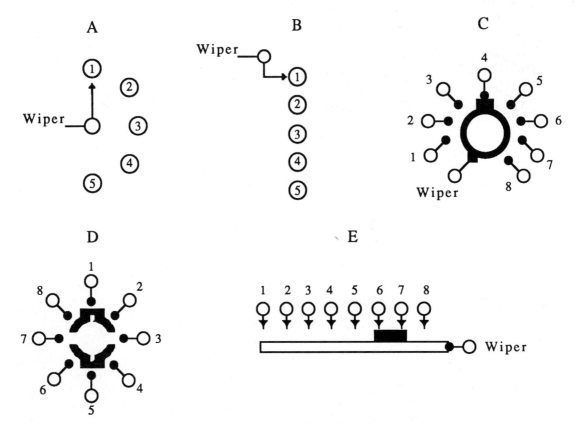

FIGURE 2-60 Symbols for wafer switches

A single-section wafer switch can be used when two or more sections of the same circuit must be switched in and out at the same time. By making some points wider on the rotating ring, certain connections can be made between them and the stationary contacts. The ring may also be broken instead of solid so that one half serves some of the stationary contacts and the other half serves the remaining contacts. For example, figure 2-60D shows a switch in which terminals 1 and 2 and terminals 5 and 6 can be closed simultaneously by rotating the shorting ring in a clockwise direction one position. Any number of connections can be made by this type of switch, depending on the construction of the shorting ring.

Another method of illustrating a wafer switch is shown in figure 2-60E. Here, the switch symbol is laid out horizontally. The bar below the row of arrowheads represents the inner shorting ring, which for illustrative purposes has been straightened out. In the actual switch, the ends of the bar are connected. As the switch is rotated, the bar moves along and makes connections to the various contacts. The principal advantage of depicting the switch in this manner is that components connected across the various contacts can be shown more easily.

Several wafer sections are often connected to a single shaft. Thus, rotating the shaft will change the connections at each section. Although most of the wafer switches shown in the symbols have 8 positions, it is very common to find wafer switches with 18 to 24 contacts.

FIGURE 2-61 Pushbutton switches

PUSHBUTTON SWITCHES

Pushbutton switches are found in almost all control panels. They are used as primary control switches or in subordinate functions such as telephone (touch-tone) dialers that begin a sequence of events leading to the main process of decoding signal pulses into actual numbers. Typical pushbutton switches are illustrated in figure 2-61.

The schematic symbols for pushbutton switches are illustrated in figure 2-62. Symbols A and B are *normally open* (*NO*) pushbutton switches. The vertical portion of the symbol represents the button. When pushed down, it moves the horizontal bar down to make a connection across the contacts, which are represented by the two circles. In symbol B, the arrowheads represent the contacts.

54 Electronic Component Symbols

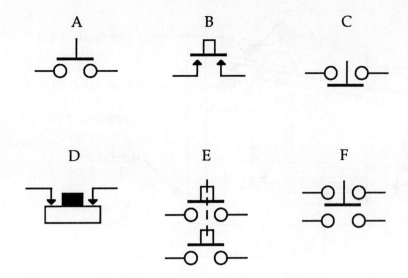

FIGURE 2-62 Symbols for pushbutton switches

If the switch is *normally closed* (*NC*), pushing the button will open the circuit. Symbols C and D represent NC switches.

A push-button switch can also be of the double-pole variety as illustrated in symbol E. Here, pushing the button closes two separate circuits. Symbol F represents a push-button switch that opens one set of contacts and closes another set of contacts when pushed.

Many other types of push-button switch combinations are possible and are common in modern electronic equipment. The symbols that represent these high-density, push-button switch configurations are combinations of the basic symbols illustrated in figure 2-62. Usually, the dashed line is used to indicate the various sections operated by a single push button.

The reference designations assigned to switches are the letters *S*, and *SW*. In addition, the letters *WS* are sometimes used to indicate wafer switches and the letters *PB* are sometimes used for push-button switches.

RELAYS

While mechanical switches are operated by mechanical movements, such as rotating or sliding a knob or pushing a button, relays are closed or opened by electromagnetic action. As current flows through the relay coil circuit, the resulting magnetic field provides the mechanical action to the contact section to complete the switching function. Relays are also able to control high-voltage circuits with lower applied voltages.

A basic relay has two separate sections: the *relay coil* section and the *relay contact* section. The relay coil section consists of a coil of wire wound around an iron core. The relay contact section completes the basic relay and is electrically isolated from the coil section. One of the contacts is movable and closes, or completes, a circuit with the stationary contact when a sufficient coil voltage is applied across the coil terminals. A typical relay is pictured in figure 2-63.

Just as there are many types of switches, there are also many types of relays available to fit a multitude of applications. These relays come in a variety of shapes and sizes and are available for basic switching applications or for special applications. A sampling of relays used in electronic equipment is shown in figure 2-64.

Symbols for relays are illustrated in figure 2-65. Symbol A represents the SPST relay. The bar at the top of the symbol is the movable contact and the arm with the arrowhead is the stationary contact. The relay coil is represented by a wire wrapped around a rectangle which represents the core. The dashed lines (sometimes omitted) signify that the core attracts the movable contact element. Symbol B is another method of representing the SPST relay. In this symbol, the rectangular bar is replaced with the symbol of an iron-core coil.

As mentioned, there are relays available other than the SPST relay represented in symbols A and B. For example, symbol C represents an SPDT relay. The movable contact completes the circuit to the upper stationary contact. This is also known as the de-energized position, in which no current flows through the coil. When sufficient current flows through the coil, the movable contact is pulled down, opening the upper contact and closing the lower one. The upper contacts are normally labeled NC (normally closed) and the lower contact is labeled NO (normally open). The movable contact arm is sometimes labeled with the letter *C* to indicate the common relay contact.

To simplify circuit layout, the connections of the relay symbol are often brought out at different sides as illustrated in symbol D.

A single relay may operate more than one movable arm at the same time. Symbol E depicts a two-section

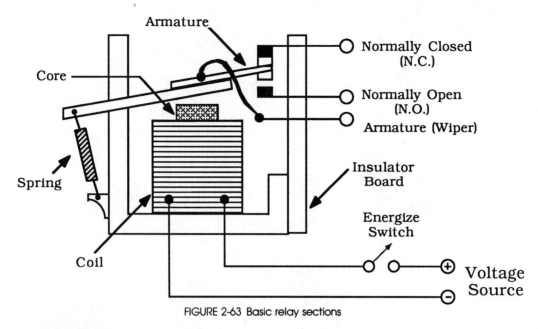

FIGURE 2-63 Basic relay sections

FIGURE 2-64 Common relay styles

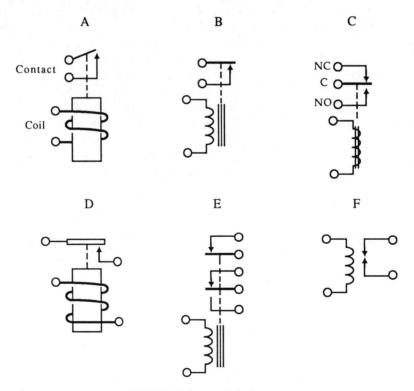

FIGURE 2-65 Relay symbols

56   Electronic Component Symbols

relay. The top portion is an SPST section that is normally closed. The bottom section is an SPDT unit that is normally connected to the upper contact.

Symbol F is a simplified style of representing an SPST relay. This symbolic style is very common in modern schematic diagrams. Although there are possibilities for innumerable relay types, the operation of a specific relay is easily determined by examining the contacts. Just remember that the relay is normally shown in its de-energized state and when energized, the bar moves *toward* the coil.

Another relay that is gaining popularity in modern electronic equipment is the reed relay (see figure 2-66). The reed relay is increasing in popularity because of its low current requirements, its increased switching speeds, and its availability in small packages. A DIP (dual in-line package) reed relay is often used in digital computer circuitry. The more common reed relay consists of two separate and independent units. These units, the coil and contact sections, can physically be removed from each other, making it possible to easily replace either of them. Normally when defective relays are found, the entire relay housing is replaced. The reed relay configuration makes it possible to replace only the defective section. These relays are also available in a number of contact types.

A more recent development is the solid-state relay. This device does not have actual contacts like the magnetic relays discussed previously. Instead, the switching device is a thyristor.

Several components are enclosed in the housing (see figure 2-67). The input is processed and connected to an LED, which glows when the conditions in the circuit to which it is connected are correct to actuate the relay. This

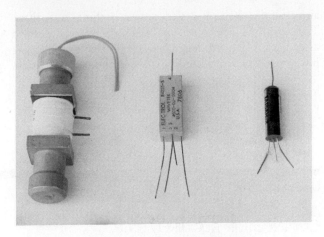

FIGURE 2-66 Reed relay packages

LED shines on a photodetector which conducts, causing the trigger current to be applied to a thyristor. Thus, the output is isolated from the input by a simple LED and photodetector device.

Other electronic devices may be included in the relay housing to prevent false triggering of the thyristor by line voltage pulses or other undesirable signals. The symbol for this type of specialized relay is represented in schematics by a drawing of the individual components within the housing and the entire unit enclosed in a solid-line or dashed-line box. Often, the black box symbol illustrated in figure 2-67 is used. That is, a square or rectangular box is used to represent the relay. The internal circuitry is eliminated and only the input and output terminals are given.

Relays are frequently separated into categories based on their applications and sizes—subminiature, miniature,

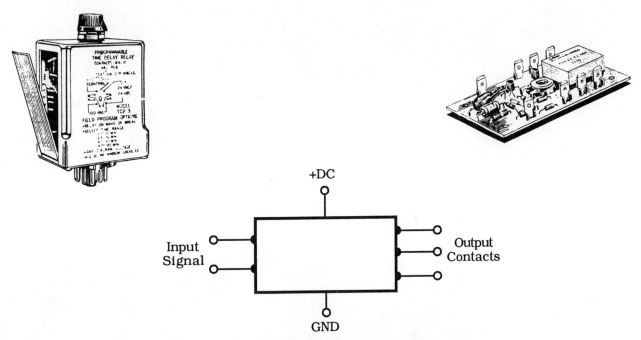

FIGURE 2-67 Solid-state relay with symbol

Graphic Symbols of Common Electronic Components

aircraft, printed-circuit, sealed, power, latching, and others. They are available in open, enclosed, or hermetically-sealed assemblies. They are designated as ac or dc and they are further subdivided according to their power operation, their number of poles, their number of contacts, their size, their coil-operating voltages, and other factors.

Reference designations for relays are *K, RE, RL, M,* and *E.*

UNIT 4 CONNECTION DEVICES

CONNECTOR REQUIREMENTS

Due to the numerous varieties of connecting devices, the problem of adopting one consistent standard for connector identification (classifying) and symbolic representation is still without adequate solution. With new connector designs constantly being added, the variety of connectors seems almost endless. These devices must meet many requirements such as size consideration, reliability, and electrical parameters. They must be designed for high-power or low-power applications, be subjected to low or high frequency signals, be adaptable for printed-circuit mounting, soldering, welding, or wire wrapping. Sockets, plugs, and jacks are only a few of the many types of connectors. All connectors have the same purpose—they provide a convenient means for connecting or disconnecting cables, circuit boards, or any other electronic interface system.

CONNECTORS

Rack and Panel Connectors

Rack and panel connectors are normally rectangular in shape and come in a great variety of contact arrangements and sizes. The MS connector, formerly known as the AN connector, was one of the first connectors specified for military electronic equipment. Rack and panel connectors are illustrated in figure 2-68.

RECTANGULAR STYLE

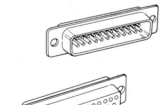

MS STYLE

FIGURE 2-68 Rack and panel connectors

58   **Electronic Component Symbols**

Miniature Connectors

Miniature connectors are normally found in such equipment as commercial or military aircrafts, space vehicles, and satellite systems. They are also utilized in large computer systems to interface subsystems to the mainframe. Miniature connectors are shown in figure 2-69.

Flexible, Flat Cable Connectors

Flexible, flat cable connectors can be used as a means of transition from a flat cable to a round-wire cable. They are found in a variety of electronic applications. Computer manufacturers are the largest users of flexible, flat cable connectors. The automobile industry has also found widespread use for this style of connector. Flexible, flat cable connectors are shown in figure 2-70.

Printed-Circuit-Card Edge Connectors

Printed-circuit-card edge connectors (see figure 2-71) enable the connection of any conventional cable, either flat or round wire, directly to the PC board. Correct PC board polarization is usually achieved by a keyed slot on the

FIGURE 2-69 Miniature connectors

FIGURE 2-70 Flexible, flat, cable connector

FIGURE 2-71 PC board edge connector

Graphic Symbols of Common Electronic Components

connector. This insures that the proper facing of the PC board is inserted into the connector.

Terminal Blocks

Terminal blocks (see figure 2-72) are used in large electronic equipment. They are used primarily to terminate wires of subassemblies that interface with the main equipment. They vary greatly in size, shape, and terminal arrangement. Individual terminals are identified by reference designations such as TB1 and TB2.

CONNECTOR SYMBOLS

Symbols for connectors are usually drawings of the actual devices.

Ac Power-Line Connectors

The graphic symbols for ac power-line plugs and sockets are shown in figure 2-73.

Symbols A and B are side views. Symbol C, which is the symbol for a plug, is an end view. Notice that the blocks in this symbol are solid. Symbol D, in which the blocks are open, represents a socket or receptacle. This system is used for most plug and socket combinations.

Combination Plug and Socket Connectors

Combination plug and socket connectors are used to provide voltage from separate power supplies to operate electronic interface units. They are also used to transfer signal voltages from one unit to another.

The symbols for combination plug and socket connectors are shown in figure 2-74. Symbol A represents the familiar plug and socket connectors found inside the chassis of many radio and TV receivers. Symbols B and C represent plug and socket connectors found mainly in power supply subunits, which provide the necessary dc voltages to operate other units. Symbols D and E represent two- and three-conductor plug and socket combinations.

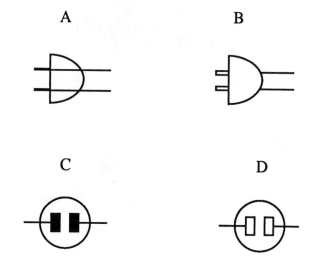

FIGURE 2-73 Symbols for ac power-line connectors

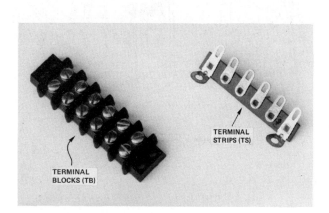

FIGURE 2-72 Terminal blocks and strips

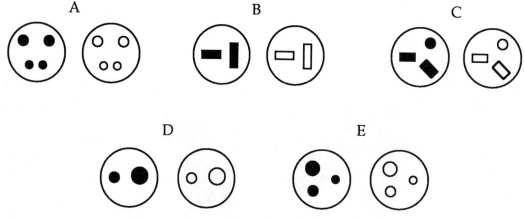

FIGURE 2-74 Plug and socket combination symbols

60   **Electronic Component Symbols**

Single-lead Connectors

Five methods of representing single-lead connectors are shown in figure 2-75. Symbols A and B are for the phono plug and socket found on the rear of many radio and TV receivers. Symbols C, D, and E depict simple one-wire connectors. The arrowhead in symbol D is the plug end and the other half of the symbol represents the socket. Symbols C and E use the same solid and hollow representations for plug and socket combinations as explained earlier.

Multi-pin Connectors

There are numerous plug and socket combinations. If the plug or socket contains so many connections that it is difficult to represent leads from all pins, a different method is sometimes used. The various points are arranged in a row and each pin and socket terminal is numbered as illustrated in figure 2-76. A separate drawing showing the pin numbering system is usually included on the schematic diagram.

Instead of the method shown in figure 2-76, individual connections may be shown as in symbols C, D, and E in figure 2-75. Each pin is then labeled with the plug or receptacle number, followed by a dash and a number identifying the pin.

Plugs and receptacles are not normally identified by code letters; however, when they are, the letters *P* or *PL* (for plug), *S* or *SK* (for socket), *X*, and *M* are frequently used. This method of representing multi-conductor connectors using the arrowhead symbol is illustrated in figure 2-77.

Audio Connectors

Audio jacks and plugs are used in many types of electronic equipment, but audio equipment manufacturers are the largest consumers of these connectors. Some common audio jacks and plugs are shown in figure 2-78.

The symbols for audio jacks and plugs are illustrated in figure 2-79. The arrow in symbol A indicates the direction in which the plug is inserted into the jack. When inserted, the tip of the plug is connected to the upper terminal of the symbol.

All of the symbols show a bar or rectangle at the front depicting that portion of the jack known as the *sleeve*. The sleeve extends out of the chassis and the plug is inserted into it. This area is sometimes shown with a heavy solid bar.

Symbol B represents a plug. When it is inserted, it strikes the V-shaped portion of the symbol connected to the upper terminal. This action disconnects the upper terminal from the center terminal. As before, the contact is made between the upper terminal and the plug.

Sometimes, the plug causes a contact to be made instead of broken. This is the case in symbol C. Many other switching combinations are possible as shown by symbols D through G.

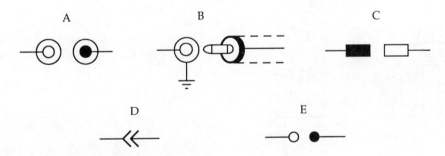

FIGURE 2-75 Symbols for single-wire connectors

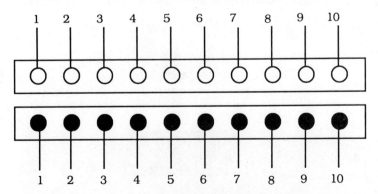

FIGURE 2-76 Alternate method of representing multi-pin connectors

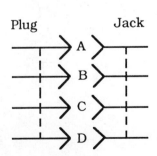

FIGURE 2-77 Representation of multi-pin conductor

FIGURE 2-78 Audio plugs and jacks

Symbols C and E, each of which has two V-shaped elements, represent a jack with two separate connectors contacting two points of the plug. The two points normally are insulated from each other. When a plug is inserted into this type of jack, one contact may be to the tip of the plug while the second contact will be to the ring which is usually located in back of the tip. A typical application for this type of plug is on stereophonic equipment where channel separation is necessary.

The X in symbol F signifies that the two points to which it is attached are joined mechanically but not electrically.

Symbols G and H represent plugs; symbol G is for a two-conductor plug and symbol H is for a three-conductor plug.

Like sockets, jacks and plugs are not usually assigned code letters. But if they are, the letter J is the most commonly used for jacks. The letters P and PL are used most often for plugs, although X is used occasionally.

Terminal Boards

Terminating devices such as terminal boards, terminal strips, or any other type of terminal blocks are essential in almost all electronic equipment. Their function is to provide a convenient means of terminating wires to complete a connection as an input or output lead from a circuit board. Input leads from antennas or output leads to speakers are places where terminal boards are employed. Terminal boards or strips are also used to complete intermediate connections where it is important to keep the wiring procedure as organized as possible. Some terminating devices are shown in figure 2-80.

FIGURE 2-79 Symbols for audio jacks and plugs

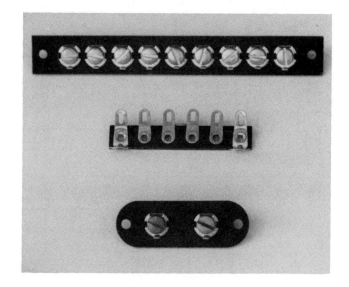

FIGURE 2-80 Terminating components

62   Electronic Component Symbols

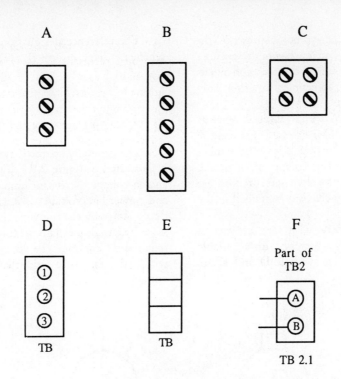

FIGURE 2-81 Terminal symbols

Symbols for terminating components are shown in figure 2-81.

Symbols A, B, and C represent terminal strips employed to terminate input or output leads such as on antennas or speaker wires. Symbols D, E, and F represent a multiple-terminal board utilized to test point positions or intermediate wire terminations. Terminal identification is generally governed by circuit convenience rather than by sequence. When necessary for simplification, parts of a given terminal board may be separated in the diagram and identified with the proper reference designation as in symbol F.

UNIT 5 SOURCES OF ELECTRICITY

DIRECT CURRENT (DC) SOURCES

Batteries are dc voltage sources that power many types of portable equipment. Essentially, all batteries produce electricity by chemical action. The interaction of an electrolyte solution with the metallic plates immersed in the solution is the basis of current production from batteries. The plates can be made of carbon, zinc, lead, nickel-cadmium, magnesium, or mercury. Regardless of their composition, the same graphic symbols are used to represent all types of batteries. Storage batteries are made up of several *cells* connected in series to increase the terminal voltage. The symbol for the basic cell and the universally-accepted battery symbols are illustrated in figure 2-82.

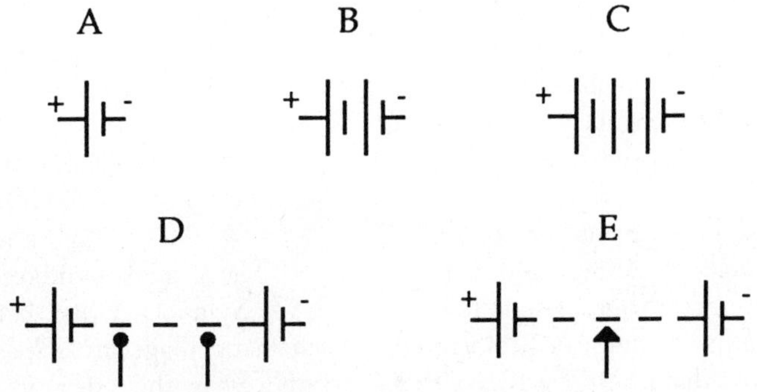

FIGURE 2-82 Cell and battery symbols

Graphic Symbols of Common Electronic Components 63

The short bar(s) in each symbol represents the negative terminal and the long bar represents the positive terminal of the cell or battery. Often the plus and minus signs are included in the symbol; sometimes, however, only the plus sign is included.

Sets of bars are added to depict multicell units as illustrated in symbols B and C. One can be led to believe that the number of pairs of bars corresponds to the number of cells in the battery. Sometimes this assumption is correct, but usually no more than four or five sets of bars are employed, no matter how many cells are contained in the battery.

Symbol D depicts a battery with taps at various points and symbol E represents a battery with a variable tap. The batteries represented by symbols D and E are very rare.

The reference designations that identify batteries on schematic diagrams are *B*, *BT*, and *E*.

ALTERNATING CURRENT (AC) SOURCES

The generally accepted symbol for ac voltage sources is illustrated in figure 2-83. This symbol can be used for any ac voltage sources including alternators, ac generators, and power-line voltages. Alternators and generators produce voltage by electromagnetic induction with the help of some external mechanical force. The external force is required to continue the rotary motion of the armature within the magnetic field that produces the alternating voltage.

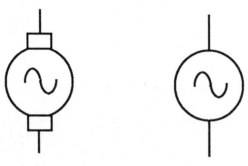

FIGURE 2-83 Symbols for ac voltage source

UNIT 6 PROTECTION DEVICES

Unless some protection is provided in electrical or electronic circuits, a short or other malfunction can destroy an entire piece of equipment. This protection is provided by *fuses* and *circuit breakers*.

FUSES

When a circuit draws too much current, the metal strip or wire inside the fuse melts and creates an open circuit which interrupts current flow. Many types, sizes, and shapes of fuses are used by the electronics industry. Some are designed to open the instant the current exceeds the rated value. Others do not open unless a constant excessive current, rather than a momentary surge, flows through them. These are known as the slow-flow type. Fuses and fuse holders are illustrated in figure 2-84.

Some fuses are enclosed in a ceramic, instead of a glass, cartridge. Fuses are rated by current and voltage values. They are available in styles from those that have leads extending from their ends so they can be soldered directly into the circuit to those that can be easily replaced by removing them from a clip or screw-type holder.

FIGURE 2-84 Fuses and fuse holders

The graphic symbols for fuses are shown in figure 2-85. Symbols A and B are used universally in modern schematic diagrams. The only difference between these symbols is the addition of the circles, which denote terminals, in symbol B. Symbol C is sometimes used for a

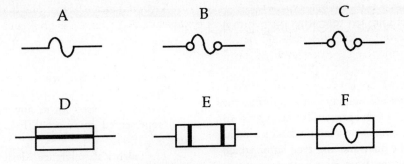

FIGURE 2-85 Fuse symbols

fuse in which a special chemical is used in place of the fuse wire.

The reference designation for fuses is the letter *F*. The electrical rating of current and/or voltage is sometimes placed beside the symbol. Occasionally, the type of fuse is also noted—for example, slow-flow type.

CIRCUIT BREAKERS

Circuit breakers perform the same electrical function as fuses, but they do not destroy themselves during the process. Instead, they open two contacts, which can be restored by pressing a button or by resetting a lever. The circuit breakers in home electrical distribution panels are a familiar application of this device. Circuit breakers found in electronic equipment are made to operate for smaller values of current than those found in home electrical systems.

Thermal Circuit Breakers

Thermal circuit breakers operate on the principle of heat. As a predetermined value of current heats a strip of metal, the metal bends causing the contacts to open. When the strip cools, the contacts can be closed by pushing a reset button.

Electromagnetically Actuated Circuit Breakers

In electromagnetically actuated circuit breakers, an electromagnetic field produced by a coil of wire has enough force to attract one of the contacts and open the circuit when the current reaches a predetermined value. Again, the circuit can be closed by pushing a reset button; however, if the overload current is still present, the contacts will remain open. In some circuit breakers, the thermal and magnetic principles are combined for improved operation.

Symbols

Several symbols are used to represent circuit breakers (see figure 2-86). Symbols A, B, C, and D denote both thermal and magnetic circuit breakers. The addition of a slanted line to the curved portion of the symbol as in symbol E indicates a switch circuit breaker. Symbols F and G represent push-pull circuit breakers.

The symbols used specifically for thermal units are H and I. They were developed by combining the basic symbol (A) with a square or two partial circles which indicate thermal action.

If the action is a magnetic one, the zigzag line shown in symbol J is used instead. A coil and a switch symbol are combined to depict a magnetic circuit breaker in symbol K.

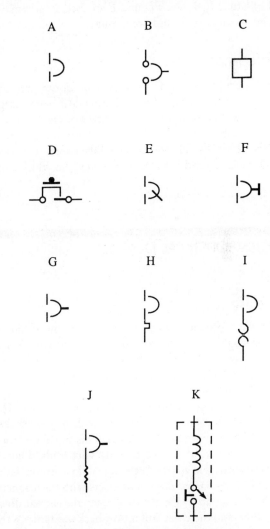

FIGURE 2-86 Symbols for circuit breakers

Graphic Symbols of Common Electronic Components

UNIT 7 LAMPS AND VISUAL SIGNALING DEVICES

LAMPS

Two types of lamps are employed for lighting the dials and other indicators in practically all types of electronic equipment: the *incandescent* lamp which is similar to the common flashlight bulb and the *neon* lamp which obtains its lighting properties from inert gases.

Incandescent Lamps

Incandescent lamps are used for illumination rather than as warning indicators. For example, many radio and TV dials have lamps behind them to make the markings visible.

The symbols for incandescent lamps are illustrated in figure 2-87. The circle around the symbol depicts the glass envelope of the bulb and the portion inside represents the filament wire which gives off light when heated.

The letters I, B, DS, PL, and E are used on schematic diagrams to identify incandescent lamps.

Neon Lamps

Neon lamps emit a soft glow when lit. For this reason, their widest application is as indicators. All neon lamps consist of two electrodes separated by neon gas.

The symbols for neon lamps are illustrated in figure 2-88. Notice that all of the symbols depict the two electrodes. The only difference is the manner in which they are drawn and the dot, which symbolizes gas, inside the envelope.

Symbol E is used for ac applications only, while the lamp represented by symbol F is used for dc applications.

The identical reference designations are used to designate neon lamps as incandescent lamps; however, the letters *NE* are also employed for neon lamps.

FIGURE 2-87 Symbols for incandescent lamps

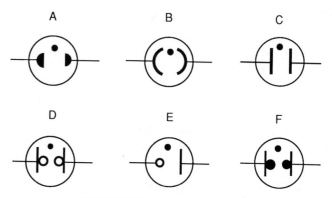

FIGURE 2-88 Symbols for neon lamps

UNIT 8 ACOUSTICAL DEVICES

SPEAKERS

Speaker systems are the final link in the chain of amplifier stages in a radio receiver, sound system, or TV receiver. The purpose of speakers is to convert the amplified electrical signals to an audible reproduction of the voice or music sounds.

All speakers operate on the following principles. The signal voltage is connected to the terminals on the speaker frame, which are known as the *basket*. This signal is carried from the basket to the voice coils by wire leads. Thus, a current that varies in step with the signal flows through the voice coil. A permanent magnet interacts with the magnetic lines of force set up by the current. When the current flows in one direction, the voice coil moves backwards along the pole piece; when the current flows in the opposite direction, the voice coil moves forward. The voice coil is held over the magnet by a fiber disc called the *spider*, which is also connected to the cone. Therefore, the cone moves as the voice coil moves, alternately expanding and compressing the air in front of it. Sound waves are nothing more than cycles of rarefied and compressed air. The sounds we hear are disturbances produced by the cone.

The size of the cone affects the range of tones that the speaker is able to reproduce. In general, a small speaker reproduces high tones, while a large speaker reproduces low tones.

Another approach to speaker design is the three-way speaker. This speaker consists of two cones fastened on a common frame. A large cone, which is located around the outside perimeter of the speaker, is used to reproduce low tones and a small cone (the *tweeter*), which is located in the center of the speaker, is used to reproduce high tones.

66 Electronic Component Symbols

Many different symbols are used to represent speakers on schematic diagrams. The most common symbols are illustrated in figure 2-89.

The acceptable reference designations for speakers are *S*, *SP*, *SPK*, and *LS*.

HEADSETS AND EARPHONES

Headsets and earphones, like speakers, convert electrical signal voltages into sound. Some earphones are constructed very much like speakers. Usually, two coils are placed over two pole pieces which are permanent magnets. These pole pieces attract a metal diaphragm that is suspended over them. As the current through the coils varies, the magnetic field it sets up is alternately added to and subtracted from the field of the permanent magnets. The changing field moves the diaphragm back and forth in step with the voltage.

Other types of headsets have crystal or ceramic elements. In these headsets, a varying voltage is applied to the crystal or ceramic slab which, in turn, moves the diaphragm back and forth.

The symbols for headsets and earphones are shown in figure 2-90. Symbols A and B are the basic symbols. Symbol C represents a double headset. Additional circles are sometimes placed inside the symbol as in symbol D. Symbols E and F designate *handsets* such as those used in telephones and some intercom systems. The addition of a line on the connecting arc in symbol F signifies a push-to-talk switch on the handset.

MICROPHONES AND OTHER SOUND TRANSDUCERS

Microphones

Microphones change varying sound waves into an electrical signal voltage that varies with the same intensity as the original sound. Before sound waves can be amplified, they must be converted into an electrical signal. After the process of signal conversion is complete, the signal can be transformed into a number of complex signals for any desired signal processing application.

The *dynamic*, or *moving-coil* microphone has a coil that moves a magnetic field, thereby converting sound into electrical waves. As the sound strikes the diaphragm, the coil movement induces a voltage in the coil.

Several other principles are used to produce electrical signals from a microphone. The *carbon microphone* consists of a brass cup filled with compressed carbon granules. A diaphragm connected to the cup is moved back and forth by the sound waves. The movement changes the pressure on the granules, thereby changing the resistance of the carbon. This resistance change in the carbon granules

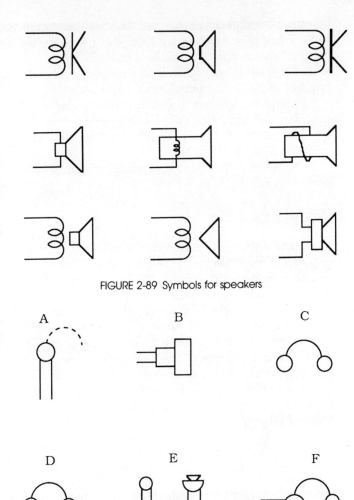

FIGURE 2-89 Symbols for speakers

FIGURE 2-90 Symbols for earphones, headphones, and headsets

results in a current that varies in step with the sound waves.

The capacitor principle is another method employed in microphone technology. The sound waves move one plate back and forth while another plate remains stationary; this causes the capacitance to change. The change in capacitance directly affects the charging and discharging of the current, which causes the sound wave variations.

Certain ceramic materials such as barium titanate generate a voltage when pressure is applied directly to the crystal surface. This phenomenon is known as the *piezoelectric effect*. When alternating pressure is applied to the crystal surface, the crystal bends or twists in synchronism with the variations. If a ceramic element is connected to a diaphragm by a drive rod, the sound pressure will bend the ceramic unit and a voltage will be generated. The resulting current is then coupled to a preamplifier for further amplification.

Some microphones employ crystals instead of ceramic elements. Crystals, however, are more susceptible to damage from humidity or high temperature.

Graphic Symbols of Common Electronic Components

Symbols for microphones are illustrated in figure 2-91. When these symbols are used, the type (crystal, capacitor, dynamic, etc.) is usually designated by a note beside the symbol. Symbols D, E, and F represent crystal, dynamic, and capacitor microphones, respectively.

Microphone symbols are often omitted from schematic diagrams because they are not an integral part of the unit. Instead, only the connector or terminal to which the speaker is connected is shown.

M, *MK*, and *MIC* are the reference designations for microphones.

Transducers

Transducers are any devices that transfer energy from one or more systems to another system. Speakers and sonar pick-up devices are transducers. While microphones are designed to pick up sounds that we hear and convert these sound waves into electrical signals, transducers are designed to respond to sounds we cannot hear. An example of this application is the use of ultrasonic sounds for remote control of a TV receiver. The application of the term transducer to ultrasonic applications is quite common.

Some of the symbols for ultrasonic transducers are shown in figure 2-92. As with microphones, many different materials are used for the construction of transducers. The type may be designated by a note beside the symbol.

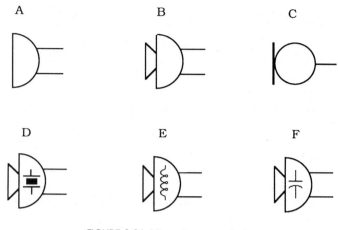

FIGURE 2-91 Microphone symbols

FIGURE 2-92 Symbols for ultrasonic transducers

UNIT 9 ROTATING MACHINERY

Generators, motors, and synchros are rotating machines. The symbols for rotating machinery are illustrated in figure 2-93. The basic circle may be used to identify generators or motors. The reference designation is placed beside the symbol.

The symbols illustrated at A are combinations of the basic circle symbol with the appropriate designation. A circle with a straight line under the designation *GEN* or *MOT* represents a dc-operated motor. The Hertz symbol under the reference designation indicates that the symbol represents an ac motor. The symbol modifiers illustrated at B are added to the basic circle symbol to identify the specific type of motor or generator. The symbols at C are simplified symbols for motors and generators.

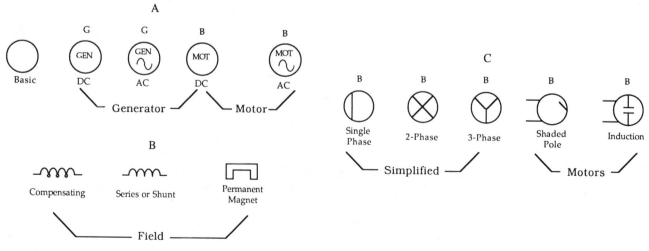

FIGURE 2-93 Symbols for rotating machinery

68 Electronic Component Symbols

UNIT 10 THE TRANSMISSION PATH

WIRES AND CABLES

All electronic components, receptacles, terminal boards, and other terminating apparatus must be connected by some means to complete electrical paths. Although printed-circuit boards provide a convenient and efficient means of connecting components, they cannot function independently without being interfaced with other components or other PC boards. Wires are the most widely used means of connection. Connections may be an actual wire between two or more points or wire (leads) extending from a component. Regardless of which method is used, wire connections are represented the same way on any schematic diagram.

A *line* denotes a wire or component lead. There are three acceptable methods of showing connection between two or more leads (see figure 2-94). In method A, the vertical line on the left intersects the horizontal line, signifying that the two wires are electrically connected. Now, notice that the lead on the right has a half circle at the crossover point. The half circle indicates that this wire bypasses the horizontal wire; therefore, the two are not electrically connected.

In method B, the dot placed at the point where the left vertical line crosses the horizontal line signifies that there is a connection at the point of intersection. A dot does not appear at the intersection of the vertical line on the right with the horizontal line so there is no connection at that point.

The systems illustrated at A and B can be confusing if inconsistencies exist on diagrams. One method must be used exclusively to avoid this problem. The most widely accepted method on modern schematic diagrams is method B.

Method C is a combination of methods A and B. The dot on the left indicates that the two lines connect. To be on the safe side, the half circle is included on the second vertical line to indicate that there is no connection at that point.

Other Methods of Denoting Connections

Electronic drafters constantly strive to make their schematic diagrams easier to read. A long, winding line is difficult to trace around a schematic when the connections are widely separated from their termination points. Eliminating as many lines as possible is one way of simplifying a schematic. For example, instead of drawing separate lines from the outputs of a multi-output power supply to their different locations, an arrangement such as that illustrated in figure 2-95A is normally used. Output voltages from the same source are indicated by a dot and are labeled at the points with their voltage values. All points connected to a source are indicated by an arrow, circle, or dot and are labeled with their voltage values as illustrated in figure 2-95B. Coded letters are used occasionally instead of listing the actual voltage values.

In figure 2-95C, the voltage output values are designated by the numbers 1, 2, and 3 in the boxes. Any point on the schematic connected to this point will have the same box and number. Other geometric shapes such as triangles and diamonds are also employed to signify connections between two points. Often, the source is

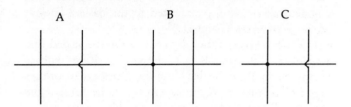

FIGURE 2-94 Wire or lead connection methods

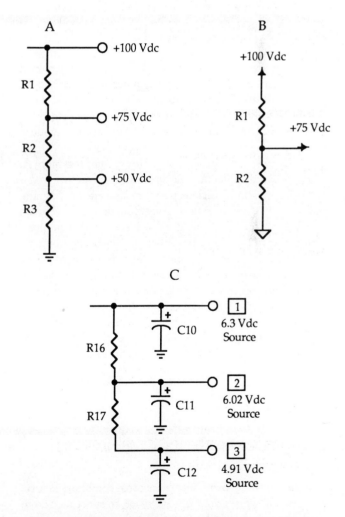

FIGURE 2-95 Method of representing multi-output voltage points

Graphic Symbols of Common Electronic Components   69

indicated by a solid symbol and the points connected to it are indicated by the outline of the same symbol.

CABLES

In many instances, two or more points are connected by cables. A cable is several single insulated wires bundled together. Cable symbols are shown in figure 2-96.

A ring on the wires represents the outer covering or insulation around the entire wire bundle. If the cable is shielded, a ground symbol may be added to the basic symbol as in symbol B. A dashed circle or a dashed line above and below the solid lines that represent the wires may also be used as illustrated in symbols C and D. A ground symbol can be connected to the dashed lines as shown in symbols C and D.

The methods used to designate a single-shielded lead are illustrated in symbols E, F, and G. The dashed lines in G may extend the entire length of the cable or wire, or they may extend for a short distance as shown here.

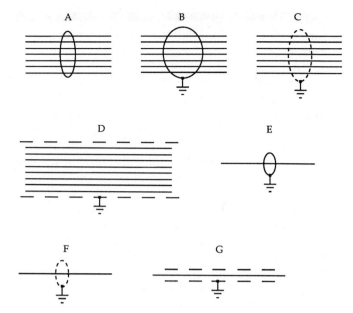

FIGURE 2-96 Cable and shielded lead symbols

UNIT 11 GROUND SYMBOLS

The term ground is one of the most misunderstood terms in electronics. This term originated in the early days of radio when radios were literally connected to the earth (ground). Another term that is used frequently for ground is *common*. The symbols used for grounds on modern schematic diagrams are shown in figure 2-97.

The most widely used symbol for ground is symbol A. This symbol also denotes earth or chassis ground. If this symbol is present at various points in a schematic diagram, there is a common return point for the entire circuit. Symbols B and C are used exclusively for this purpose. Otherwise, a line must be drawn to all of the common points in the circuit. All points exhibiting this symbol are considered to be connected electrically. This method of representing several common points on a complex schematic greatly reduces clutter, making it easier to interpret the diagram.

Often, the ground symbol signifies the chassis of the equipment. Instead of the various points being connected by a wire, they are connected to the metal chassis. The metal chassis, therefore, serves as the common conductor for all of the points connected to it. This practice is most common on ac/dc equipment in which one side of the line voltage is connected directly to the chassis. This connection makes the chassis hot where a potential shock hazard might exist. In this type of equipment, the chassis is designated by symbol D.

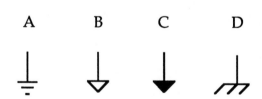

FIGURE 2-97 Ground and chassis symbols

UNIT 12 COMPONENT COMBINATIONS

Two or more components are often combined to form a single unit. This is done to save space, to cut costs, and to eliminate any electrical interaction between components.

COIL AND TRANSFORMER COMBINATIONS

Many combinations of components are possible. For example, a coil may be wound directly over the body of a

70   Electronic Component Symbols

resistor to form a parallel resistor-inductor (parallel RL) combination. Another common practice is to connect a capacitor across one or both windings of a transformer.

Actual component combinations along with their schematic representations are illustrated in figure 2-98. The symbols may be enclosed in a dashed-line box which signifies that the components within the box are part of an individual combination unit. Notice the ground symbol connected to the box for the transformer-capacitor combination at B. It indicates that the components within the box are *shielded* from the rest of the circuit. That is, stray magnetic fields cannot enter or leave the shielded enclosure. This shielded box is normally connected to the chassis and is, therefore, grounded to it.

Some other component combinations are illustrated in figure 2-99. The units represented by symbols A, B, C, and D contain a coil and a capacitor. Symbol A has a fixed capacitor connected directly across a coil. In symbol B, the coil is also parallel to the capacitor; however, the capacitor in this combination is variable. The coils and capacitors in symbols C and D are connected in series.

The transformer in symbol E has a resistor connected across one winding and a capacitor connected across the other. Many other similar combinations are possible.

Symbol F represents a transformer, three capacitors, a coil, and a diode in a single enclosure. This symbol also depicts the diode being mounted in a separate housing which is shielded from the main enclosure.

The dashed line around any component combination with the ground symbol connected to it is used to designate a shielded enclosure. As was mentioned previously, when the dashed lines are drawn around a lead or a group of leads, they represent a shielded wire or cable. At other times, dashed lines may be placed around a tube symbol to indicate that a shield is placed over the tube after it has been inserted into the socket.

PACKAGED ELECTRONIC NETWORKS

Packaged electronic networks (see figure 2-100) contain several components connected in a specific configuration. These networks contain a variety of combinations of resistors, capacitors, and coils, which are bound to a base plate and sealed with a protective ceramic or plastic coating. Such units are extremely resistant to moisture, temperature, and shock.

Packaged electronic networks are usually represented schematically by using regular resistor, capacitor, or coil

(A) RESISTOR AND COIL

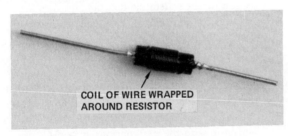

(B) TRANSFORMER AND CAPACITOR

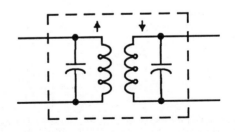

FIGURE 2-98 Component combinations

Graphic Symbols of Common Electronic Components

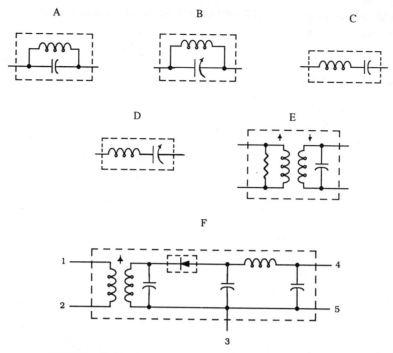

FIGURE 2-99 Coil, capacitor, and transformer combinations

FIGURE 2-100 Packaged electronic networks

symbols. The combination is then enclosed with dashed lines to indicate that it is contained in a single unit. Each lead is numbered at the point where it extends from the dashed lines. Some of the networks found on modern schematic diagrams are shown in figure 2-101.

The dashed lines around the component network are sometimes omitted. Symbol D is used occasionally instead of showing the internal connections. The symbol may be enclosed by the dashed lines. The internal connections are usually shown elsewhere as a footnote or in a separate section of the schematic diagram.

There are several methods of assigning reference designations to these units. Some companies assign the letters *A*, *E*, *K*, *M*, *N*, *X*, *PC*, *PN*, *DC*, or *RC*. Other companies assign the letters *R* and *C* to the unit and designate the individual components. Still others combine the two methods, assigning a letter and a number to the entire unit and then designating the individual components within it with the designations *R* and *C*.

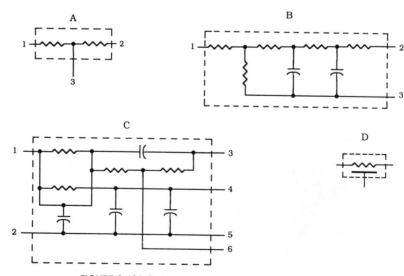

FIGURE 2-101 Packaged electronic network symbols

72 Electronic Component Symbols

UNIT 13 LOGIC SYMBOLS

A *logic symbol* is a graphic representation in diagram form of a logic function. The basic logic functions are the *AND, NAND, OR, NOR,* and *NOT* or *NEGATION*. At times, the actual circuit configuration is not of as much interest as the logic function being performed. For example, the actual circuit within an integrated circuit which performs the OR function (see figure 102) is seldom of interest.

Uniformly shaped symbols such as rectangles may be used in logic diagrams or distinctly shaped symbols may be used for the most common functions and rectangles for others. In either case, internal abbreviations identify the function when it is in rectangular form. Military logic diagrams require the use of distinctly shaped symbols.

THE AND FUNCTION

The AND function can be explained simply using two switches connected in series (see figure 103). If switch A or B is open, a complete path is not provided through the circuit; consequently, there is no output signal. However, if switches A *and* B are closed, the circuit is complete and there is an output. If more switches are added in series, all of the switches must be closed to make a complete path for the output signal. These switches represent inputs. The AND function is performed by an AND gate in which all of the inputs must be pulsed by a signal to obtain an output signal.

The AND gate is usually represented by symbol B. The two lines extending from the left labeled A and B are the inputs. In symbol C, a rectangular or square box is used to represent the logic gate. A qualifying symbol (letter or other character) indicates the type of gate. Here the ampersand (&) signifies the AND function.

Table D shows the various combinations of inputs and outputs. This tabular device is called the *truth table*. The zero represents the open position of the switch (no input signal) and the one represents the closed position (signal applied). The one is commonly referred to as the *logic high* while the zero is referred to as *logic low*. The resultant output conditions can be determined by evaluating the truth table. For the AND function, therefore, it can be seen that both inputs A and B must have signals applied to produce an output signal (logic high). Any other combinations of input will result in a zero output condition (logic low).

THE OR FUNCTION

If two switches are connected and if either switch A *or* switch B is closed, there will be an output (see figure 2-104A). This logic function is the OR function which is performed by an OR gate. More switches may be added to represent the input to this gate. Symbols B and C represent the OR gate.

The truth table for the OR function is illustrated at 2-104D. As shown in the truth table, if either input is a one or if both inputs are ones, the output will be a one (high). The only condition that will cause the output to be at zero (low) is when both inputs are at the zero, or low, state.

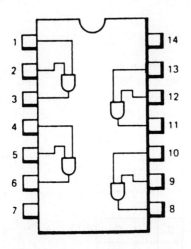

V_{cc} = Pin 14, GND = Pin 7

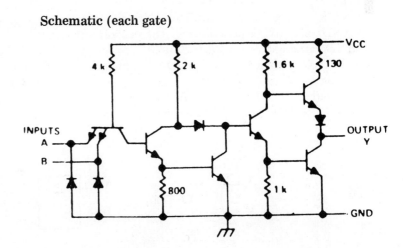

FIGURE 2-102 Circuitry that performs an AND function

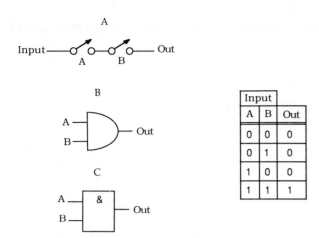

FIGURE 2-103 The AND function with logic gate symbols

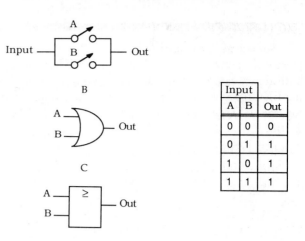

FIGURE 2-104 The OR function with logic gate symbols

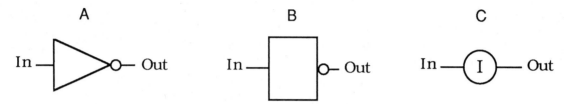

FIGURE 2-105 Logic symbols for the NOT function

THE NOT AND NAND FUNCTIONS

In logic terminology, the NOT function signifies negation or reverse of a condition. For example, if one position of a two-position switch is designated A, the other position is $NOT\ A$, which is written as $\overline{A}$ or A^1. An *inverter* reverses the normal condition of a circuit.

The logic symbols for the NOT function are illustrated in figure 2-105. In symbol A, the basic amplifier symbol (the triangle) is slightly modified by the circle which is placed at the output of the symbol. Symbol B illustrates the rectangular symbol for the inverter. Symbol C is another way of representing the NOT or inversion function.

If an inverter is combined with an AND circuit, a NOT-AND, or NAND gate, is formed. The symbols for the NAND gate are shown in figure 2-106. In symbols A and B, the circle combined with the AND symbol signifies the negation or inversion function.

The truth table at C illustrates the logic function of the NAND gate. Notice that if either or both of the inputs

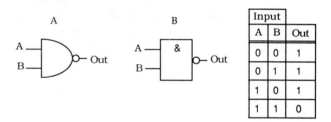

FIGURE 2-106 The NAND function with logic symbols

is a zero, the output will be one and if both inputs are one, the output will be zero.

THE NOR FUNCTION

When the NOT function is combined with the OR function, the NOT-OR, or NOR function, is created. The NOR gate performs this logic function. Symbols A and B in figure 2-107 represent the NOR gate. The truth table is shown at C.

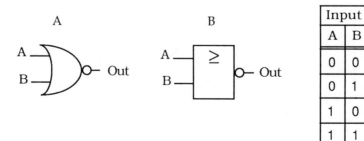

FIGURE 2-107 The NOR function with logic symbols

74   Electronic Component Symbols

EXCLUSIVE OR FUNCTION

Sometimes a circuit is required in which an output is present if one switch or the other is activated, but not when both are activated. This logic function can be performed by the EXCLUSIVE OR gate. The symbols and truth table for the EXCLUSIVE OR gate are shown in figure 2-108.

OTHER LOGIC SYMBOLS

Other logic symbols may be encountered on schematic diagrams, but usually they will differ only slightly from the basic symbols described. For example, while only two inputs were shown for each function, any number of inputs may be used. In symbols A and B in figure 2-109, the lines of the AND and OR gates are extended to accommodate more inputs. If space does not allow all of the inputs to be shown, the shape of the rectangular symbol is altered instead of using extensions.

Symbol C may be used to represent the AND gate and symbol D can be used for the OR gate.

Symbols E and F represent an *oscillator*. The output of this device is a uniform repetitive signal which alternates between the zero and one state. The G inside the rectangular symbol stands for generator.

Often, it is desirable to delay a signal at a certain point in the circuit. Such delays are necessary because any signal processing requires a certain amount of time known as the *set-up time*. Thus, if it is desired that a processed signal arrive at a given point at the same instant as an unprocessed signal, the unprocessed signal must be delayed before being applied to the same point. Symbols G and H identify delay elements. The two vertical lines in each symbol represent the input side. The output from this element will assume its one state only after a specified period of time following the transition of the input to the one state. The output will revert to the zero state only after the specified period of time following the reverting of the input to the zero state. The amount of delay time is usually indicated on the symbol. Delay elements can also be made adjustable; in this case, an arrow is added as shown in symbol I.

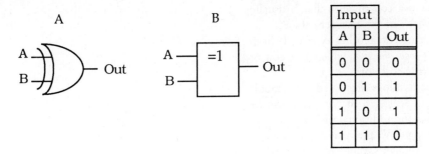

FIGURE 2-108 The exclusive-OR function with logic symbols

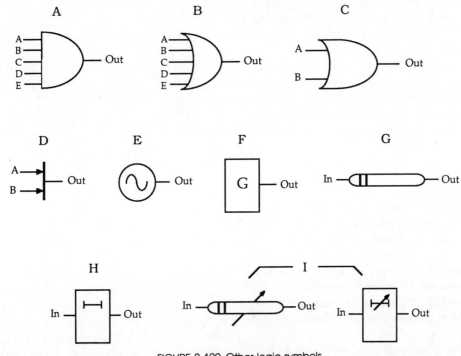

FIGURE 2-109 Other logic symbols

Graphic Symbols of Common Electronic Components 75

UNIT 14 MISCELLANEOUS SYMBOLS

ANTENNAS

Antennas are a vital link between transmitting and receiving equipment. At the transmitter, the antenna is the final unit through which the electromagnetic radio waves pass to be sent through the air. At the receiver, the antenna is the unit that intercepts the transmitted radio waves and conveys, or couples, them to the tuned circuits that make up the front end on any receiver system. Sometimes the antenna is not part of the main equipment; it is mounted externally away from it. In such a case, the antenna symbol may not be included on the schematic diagram. Instead, only the terminal to which the antenna is connected may be shown.

Symbols for antennas are shown in figure 2-110. Symbols A, B, C, and D generally represent *external antennas*. Symbols E, F, and G represent the familiar *loop antennas*. A loop antenna is a coil of wire that is usually fastened flat against the back of the receiver cabinet.

Symbols H and I represent another type of built-in antenna called the *ferrite-loop antenna*, which is a coil of wire wound around a ferrite-core form.

Some antennas, such as those illustrated by symbols H, I, and J, can be tuned. A length of wire may also be attached to these antennas for additional pickup.

Methods of indicating TV antennas are shown in symbols K through O. Although symbol K is used for all types of TV antennas, it actually represents a *dipole* antenna, which is a basic TV antenna. Symbol L is another basic TV antenna, the *folded dipole*. Symbols M, N, and O represent a *monopole* antenna, which is a single telescoping rod built into portable TV receivers. Two of these symbols may be used to indicate two telescoping elements.

Loop and ferrite-core antennas are shown in figure 2-111.

Since all antennas are so much like the basic coil (the inductor), the letter *L* is used as the reference designation. The letters *E* and *M* are also used.

CRYSTALS

Crystals are made from materials such as quartz that have the unique property of generating a voltage when pressure is applied to them (the piezoelectric effect). When an alternating voltage is applied to the crystal, the crystalline structure bends or twists in synchronism with the voltage variation. By cutting the crystal at various angles and dimensions and by placing a metallic plate on each side of the crystal, the crystal can be made to oscillate or vibrate at a specific frequency which is known as the natural *resonant frequency*.

Once the crystal starts oscillating, only a very small signal voltage *at the same frequency* is required to obtain

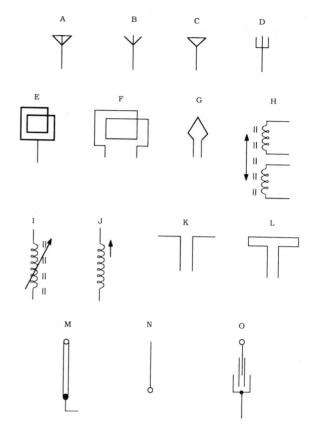

FIGURE 2-110 Antenna symbols

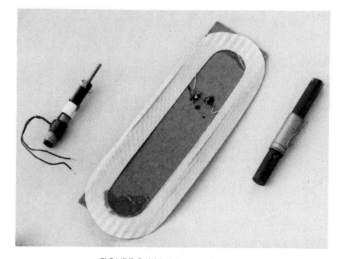

FIGURE 2-111 Loop antennas

large amplitude oscillations from the crystal. These oscillations of alternating voltage are often connected to the input circuit of a vacuum tube or semiconductor device. Since the crystal will oscillate at only one frequency, the frequency of the crystal oscillator stage will remain relatively constant. Crystals are also used as *filters* for tuning amplifiers. As filters they allow only a specific band of frequencies to pass while rejecting other frequencies.

The symbols for crystals and some illustrations of crystal housings are shown in figure 2-112. The two horizontal bars in each symbol represent the metal holders and the rectangular or slanted lines represent the crystal element.

The letters Y and X are used to identify crystals on schematic diagrams.

CRYSTAL HOUSINGS

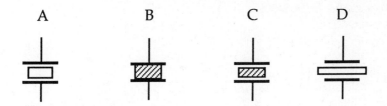

FIGURE 2-112 Symbols for crystals

SECTION 3
ASSIGNMENTS

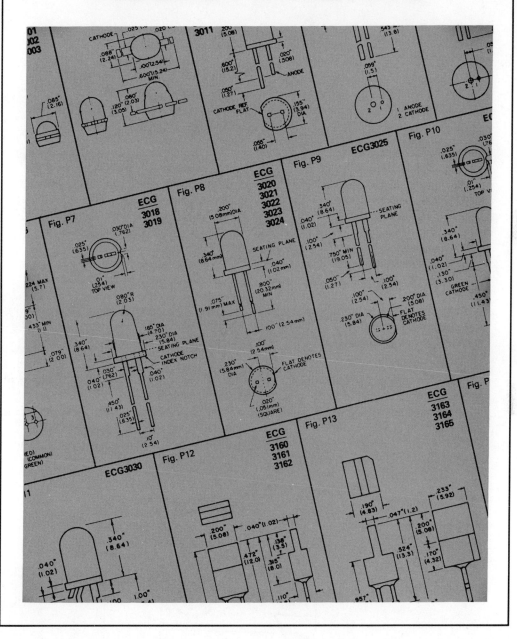

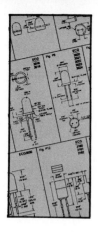

Assignment 1: Familiarization with Basic Electronic Hand Tools

PURPOSE

- To identify the basic electronic hand tools by their correct names.
- To develop safe usage of each hand tool for specific applications.
- To manipulate the proper tools to produce common electrical wire splices.

COMPONENTS & EQUIPMENT

Components

- Wire set (miscellaneous hook-up wire, coaxial cable and twisted-shielded pair)

Equipment

- Wire strippers
- Diagonal cutting pliers
- Needle nose pliers

EXERCISE 1: WIRE STRIPPING

Stripping wire is the process of removing the insulating material from a wire. Insulating materials include vinyl plastic (polyvinyl chloride), Teflon (polytetrafluoroethylene), rubber, cloth, and enamel.

In this exercise, you will develop the technique of stripping various types and gauges of wires used in the electronics industry.

When stripping the insulation from wire leads, care must be exercised to prevent damage to the conductor. The insulation should be stripped clean with no nicks in the wire.

Approved strippers must be used. (Strippers are normally approved by line supervisors or by individuals responsible for quality assurance.) Use thermal insulation strippers where applicable—for stripping Teflon wire or in confining areas where the pulling action of stripping is virtually impossible (see figure 3-1).

FIGURE 3-1 Thermal strippers

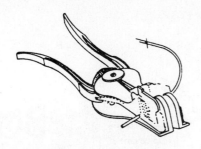
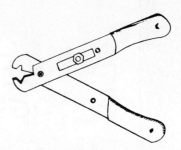

PRECISION CUTTING STRIPPERS   ADJUSTABLE STRIPPERS

FIGURE 3-2 Precision and adjustable cutting strippers

For stripping glass braid, use precision cutting strippers. These may also be used in lieu of thermal strippers for removal of other types of insulating material (see figure 3-2).

Adjustable cutting strippers are the most economical strippers one can purchase, but they are not recommended for use by industry because they have a fixed gauge setting which makes it difficult to strip different gauges of wire. With a great deal of practice, however, one may develop wire stripping skills using them. It is the primary objective of this exercise to develop the feel of stripping various gauges of wire without any adjustment control on the stripping tool.

Procedure

Separate the solid conductor wires from the wire set. Using the adjustable cutting strippers, practice stripping insulation on these solid wires by stripping approximately 0.5″ ± 0.125″ (1.27 cm) from both ends of each solid wire.

NOTE: *Adjust the depth adjustment screw on your stripping tool to close the cutting edges (see figure 3-3).*

Check for nicks in the conductor after completing each stripping action. Practice stripping on the same solid wires until you are satisfied with your performance.

Repeat the stripping procedure on stranded wires. Stranded wires are more difficult to strip due to the multiconductor strands that make up the conductor. With practice, however, you should be able to develop the feel to control the squeeze.

Broken strands should not exceed the numbers shown in table 3-1.

FIGURE 3-3 Adjusting depth of cut

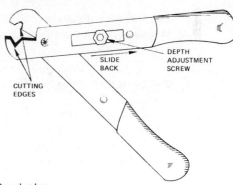

TABLE 3-1 Number of Broken Strands Allowed in a Stranded Conductor

STRANDS IN CONDUCTOR	BROKEN STRANDS ALLOWED
8–15	1
16–15	2
19–25	3
26–36	4
37–40	5
41 & up	6

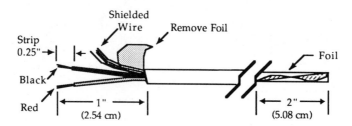

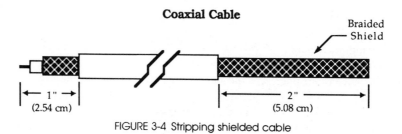

FIGURE 3-4 Stripping shielded cable

Examine the coaxial (RG-58) and shielded-pair cables and note the similarities and differences of the following elements:

- Color of outer insulation
- Insulation material
- Internal shield conductor
- Internal center conductor(s)

Using a stripper, remove approximately 1" ± 0.125" (2.54 cm) from one end of the shielded-pair cable (gray insulation). Remove the foil from this end and strip approximately 0.25" ± 0.125" (0.635 cm) from the end of each conductor. Twist the shielded wire to prevent stray ends.

Strip approximately 2" ± 0.125" (5.08 cm) from the opposite end of this cable. *Do not remove* foil from this end (see figure 3-4).

Remove approximately 1" ± 0.125" (2.54 cm) of insulation from one end of the coaxial cable (RG-58) and approximately 0.2" ± 0.125" (5.08 cm) from the opposite end. After removing the outer insulation from both cables, inspect the braid or foil for any severe nicks or severed braid wire.

EXERCISE 2: COMMON ELECTRICAL WIRE SPLICES

Splices can be described as a means by which two or more wires can be connected electrically without a stationary terminating point. The primary objective of this exercise is to produce three common electrical splices—the rat-tail splice, the tee splice, and the western-union splice. A secondary goal of this exercise is to develop skill in manipulating the needle nose pliers to produce the required twists and forms.

Procedure

Using diagonal cutters, measure and cut six 4" ± 0.125" (10 cm) pieces of solid wire.

The Rat-tail Splice. Strip approximately 1" ± 0.125" (2.54 cm) of insulation from two pieces of solid wire prepared in the previous step. Twist the two conductors finger tight, approximately three turns. (See figure 3-5 for the proper sequence for producing the rat-tail splice.) Using the needle nose pliers, complete the twisting action to insure a tight union between the two conductors. Use diagonal cutters to cut loose ends from the spliced ends.

Step 1: Strip wires

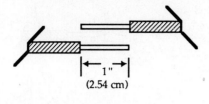

Step 2: Initial twist

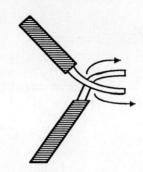

Step 3: Final twists and snip ends

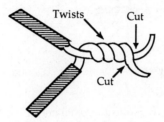

FIGURE 3-5 Rat-tail splice

The Tee Splice. Using wire strippers, remove approximately 0.75" ± 0.05" (1.9 cm) of insulation from one end of a strip of 4" solid wire prepared earlier. (See figure 3-6 for the preparation of the second wire before the splice is made.) It is suggested that diagonal cutters be used to remove the insulation around the midportion of this second wire.

Complete the tee splice as illustrated in figure 3-6.

Step 1: Strip wires

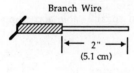

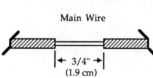

Step 2: Initial twist

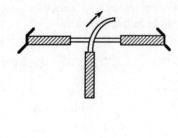

Step 3: Final twists and snip ends

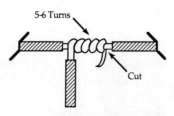

FIGURE 3-6 Tee splice

Assignment 1   83

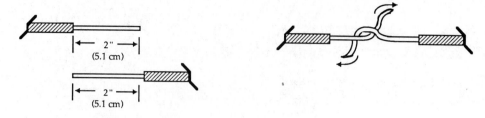

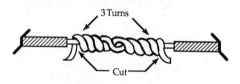

FIGURE 3-7 Western Union splice

The Western-Union Splice. Remove approximately 1" ± 0.125" (2.54 cm) of insulation from the last two pieces of 4" solid wire. Following the illustrations in figure 3-7, produce the Western-Union splice.

QA CHECKLIST: ASSIGNMENT 1

Exercise 1: Wire Stripping

Check for:

1. Nicks on stranded wire
2. Proper removal of insulation
3. Broken strands in stranded wire
4. Proper removal of insulation around TSP and coaxial cable according to required dimensions
5. Nicks in braid or foil

Exercise 2: Common Electrical Wire Splices

Check for:

1. Proper removal of insulation according to required dimensions
2. Uniformity of twist for rat-tail splice
3. Uniformity of turns around base wire for tee splice
4. Uniformity of turns around both wires for Western-Union splice
5. Proper distance between spliced wire and insulation

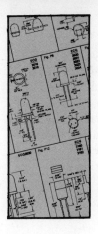

Assignment 2: Soldering Wire Terminators

PURPOSE
- To correctly tin stranded wire before terminating to a junction point.
- To solder the wire splices prepared in Assignment 1.
- To develop skill in soldering through controlled soldering exercises.

COMPONENT & EQUIPMENT

Components
- Wire set (collection of solid and stranded wire)
- Splice samples—rat-tail, tee, and Western-Union—from Assignment 1
- Solder-pattern, printed-circuit board

Equipment
- Soldering station
- Tool box
- Printed-circuit board soldering holder (see figure 3-8)

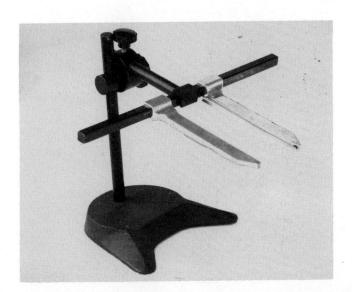

FIGURE 3-8 PCB holders

85

EXERCISE 1: TINNING STRANDED WIRE

Difficult soldering problems, such as *solder wicking, insulation damage,* and *frayed ends,* can develop when making terminations on stranded wire. These types of problems make it difficult to wrap the conductor around a terminal post or other terminating points.

Soldered connections are normally made by initially pretinning the leads or parts (if necessary) that are to be soldered together. The advantages of pretinning are quick, easy solder flow and control of the amount of solder. With stranded wire, the wire is tinned to bond the strands of the conductor together to prevent fraying during the crimping procedure which is performed before soldering. Crimping will be discussed in detail in the following assignment.

Procedure

Separate the stranded wires that were used in Assignment 1. Use *only* the stranded wires that are not tinned (pretinned stranded wire is available).

Tinning stranded wire is a relatively simple process when the proper soldering equipment and the correct technique of feeding solder to the wire are used. Strip approximately 0.5" ± 0.125" (1.27 cm) of insulation from each end of the stranded wire samples.

The tinning procedure can be completed in four steps (see figure 3-9):

1. Twist the stripped end finger tight.
2. Fasten the alligator clip on the holder.
3. Apply a small amount of solder for a bridge effect.
4. Heat the wire and feed the solder simultaneously. Draw the tip towards the end of the wire.

After tinning the wires, inspect each wire for:

- Excessive solder
- Frayed strands
- Solder wicking
- Insulation damage

Step 1: Twist strip end.

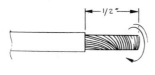

Step 2: Fasten Wire

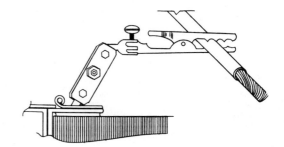

Step 3: Tip preparation

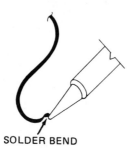

Step 4: Tinning procedure

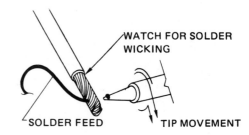

FIGURE 3-9 Tinning the stranded wire

EXERCISE 2: SOLDERING WIRE SPLICE SAMPLES

The procedure for soldering wire splices is similar to that for tinning stranded wire. The electrical union made by the splice is a strong mechanical connection developed by the style of wrap characteristic of the splice type. Although a well made splice can produce a good connection, the connection can be made better by soldering.

Procedure

The rat-tail splice will be soldered first. Before attempting to solder the splice sample, review the tinning procedure illustrated in figure 3-9.

When you are ready to begin soldering, place the splice sample in a stationary holder. Clean the soldering iron tip on the damp sponge and apply a small amount of solder to the tip to create the bridge effect. Feed the solder to the joint, not to the tip, making sure that the movement of the tip and the solder is a smooth, simultaneous action to the end of the splice joint.

Inspect the soldered sample for adequate solder coverage and shiny appearance. If too much solder was applied, a solder sucker or a solder wick can be used to remove excess solder. The solder sucker is a quick and effective means of removing excessive amounts of solder. When working with this or any other method of removing solder, it is important to work *quickly* and *effectively*: quickly, so as not to burn the insulation and effectively, so a second attempt is not required.

The solder wick is a special type of braided wire saturated with a flux material. It removes solder from a connection by drawing the solder onto the braided wire.

If too little solder was applied, reheat the joint and apply more solder. Be careful not to burn or melt the insulation.

If the soldered joint exhibits a dull luster, it is probably a cold-soldered joint. Remove the solder and resolder the joint.

The tee splice or the Western-Union splice will be soldered next (see figure 3-10 for an illustration of the soldering procedure). After completing the soldering procedure, inspect each soldered sample.

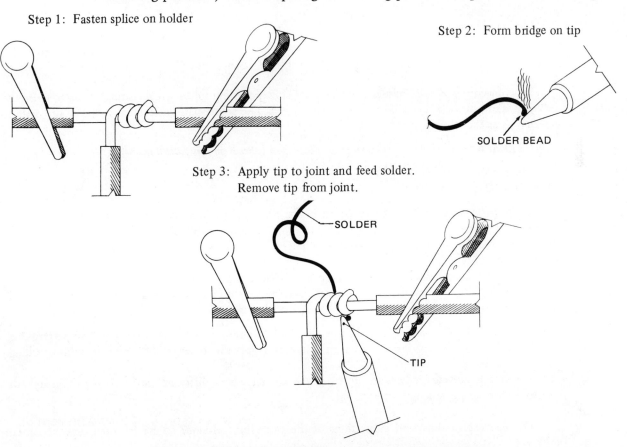

FIGURE 3-10 Soldering technique for tee splice or Western Union splice

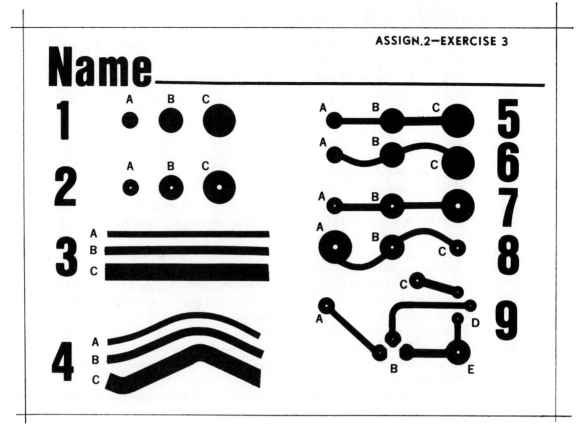

FIGURE 3-11 Solder pattern PCB

EXERCISE 3: DEVELOPING SOLDER FEEDING SKILL

The objective of this exercise is to develop the skill of feeding solder while heating the patterns on the solder pattern, printed-circuit board (see figure 3-11). The solder-pattern, printed-circuit board will be referred to as the pattern board.

NOTE: *It is not necessary to drill any of the donut-pad patterns on this pattern board.*

Procedure

Write your name on the pattern board in the appropriate place. A quick inspection of the board reveals a variety of patterns—different sized dots, donut pads, straight and curved lines, and combinations of dot/donut pads with straight and curved lines. Each pattern is numbered and lettered for easy identification of which soldering process to perform.

Fasten the pattern board on the PCB (printed-circuit board) holder.

Following are the terms used to identify the patterns on the pattern board and the soldering technique to be used for each pattern:

- All of the solid dots are identified as beads. Apply a lump, or bead, of solder on these patterns.
- All of the donut-pad patterns are identified as such. Apply the solder on these patterns without filling in the holes.
- The straight and curved lines are identified as such. They have different widths. Apply a thin layer of solder to the entire length of each line.

Detailed explanations and illustrations of the soldering technique to use for each type of pattern follow.

Dot Patterns. Applying the beads of solder to the dot patterns can be easily accomplished by feeding the correct amount of solder while heating the pattern with the soldering iron tip. When the right amount of solder has been applied, draw the solder and the tip away at the same time.

CAUTION: Excessive heat on the printed-circuit patterns may cause the pattern to lift from the board. Work quickly and effectively.

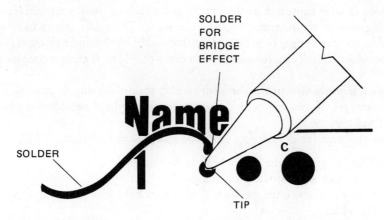

Donut-pad Patterns. Use the same soldering technique for the donut-pad patterns as for the dot patterns, but control the solder feed to avoid filling the holes with solder.

Line Patterns (straight or curved). Apply a minimal amount of solder to the line patterns. The solder, which is heated by the tip, should travel with the tip as it is fed to the trace.

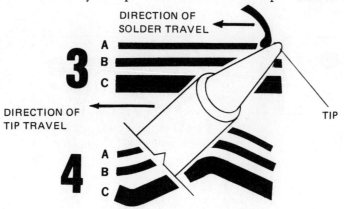

Soldering Sequence. Following is the suggested sequence for soldering the pattern board. Solder the beads (1 A, B, C), the donut pads (2 A, B, C), the straight line patterns (3 A, B, C), and the curved line patterns (4 A, B, C).

NOTE: Remember to use a minimal amount of solder on the lines (traces).

Solder pattern 5, which consists of beads A, B, and C connected by a straight run (trace), next. The straight run must be coated with a thin layer of solder, while beads A, B, and C require beads of solder as in pattern 1.

Pattern 6 is identical to pattern 5 except that it has curved runs. Use the same soldering procedure as was used for pattern 5.

Pattern 7 is similar to pattern 5 except that it has donut pads instead of the dots (beads). Use a small amount of solder on the straight runs which connect donut pads A, B, and C and use the same soldering procedure as was used for pattern 2 on the donut pads.

Pattern 8 is similar to pattern 7 except that it has curved runs. Use the same soldering procedure as was used for pattern 7.

Solder pattern 9, which is a common printed-circuit layout of an electrical circuit, last. Use the same soldering procedures as were explained previously to solder each of the pattern elements.

After soldering each of the patterns, inspect your work. The beads should have smooth, rounded tops; there should be no rough edges or peaks. The donut-pad holes should be clean. If they are filled with solder, remove the solder. Check also for excessive amounts of solder. If there is too much solder, remove some of it.

The solder wick is just as effective as the solder sucker for desoldering, or removing solder, but each method has its advantages. Normally, the solder sucker is used for large desoldering jobs while the solder wick is used for small jobs.

Following are the steps for desoldering using the solder wick (see figure 3-12):

1. Place the end of the wick on the solder bead to be removed.
2. Heat the soldered pad and the wick simultaneously.
3. When solder begins to flow onto the wick, feed the wick into the joint and watch the solder being drawn into the wick.
4. Snip the wick with a diagonal cutter and repeat step one if necessary to remove more solder.

QA CHECKLIST: ASSIGNMENT 2

Exercise 1: Tinning Stranded Wire

Check for:

1. Solder wicking
2. Frayed strands
3. Excessive solder on wire
4. Insulation damage

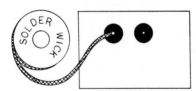

Step 1: Apply wick

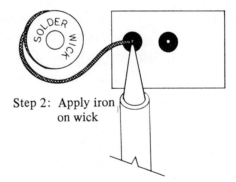

Step 2: Apply iron on wick

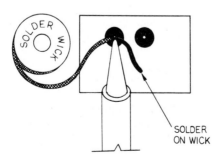

Step 3: Feed wick to joint

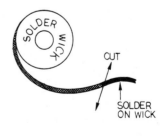

Step 4: Snip end of wick saturated with solder

FIGURE 3-12 Removing solder with solder wick

Exercise 2: Soldering Wire Splice Samples

Check for:

1. Excessive solder coverage on splice
2. Insufficient solder coverage on splice
3. Cold-soldered joint
4. Solder creep
5. Insulation damage

Exercise 3: Developing the Solder Feeding Technique

Check for:

1. Excessive solder on patterns
2. Insufficient coverage
3. Cold-soldered patterns
4. Excessive heat on patterns
5. Uniformity of solder coverage on line patterns

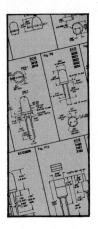

Assignment 3: Connecting Wire and Component Leads to Terminal Strips

PURPOSE

- To develop the technique of crimping wire or component leads on terminal strips and turret terminals.
- To develop more control of the basic hand tools for use in the wire crimping procedure.
- To interpret wiring instructions from a wire list.
- To develop the technique of soldering wire or component leads on turret terminals and terminal strips.

COMPONENTS & EQUIPMENT

Components

- Wire set (same set as used in previous assignments)
- Assortment of resistors (two ¼-watt carbon resistors, three ½-watt carbon resistors, two 1-watt carbon resistors, one 2-watt carbon resistor)

Equipment

- Soldering station and solder
- Basic hand tools (diagonal cutters, needle nose pliers, wire strippers, solder sucker, soldering station)
- Printed-circuit board holder
- Pre-assembled terminal board (see figure 3-13)

EXERCISE 1: WIRE CRIMPING AND COMPONENT MOUNTING

This exercise is designed to develop skill in wire crimping and component mounting. Electronic components must be mounted with care to assure uniformity and neatness. Consideration must be given to the variations of component mounting methods, the materials available, and whether the components will be interfacing with mother boards, circuit panels, terminal boards, or connectors. When connecting wires or component leads to terminating points, the crimp or wire wrap should be sufficient to hold the wire in place before soldering. Leads and wires should be connected mechanically to the terminals before soldering. This will prevent motion between the connections. The soldered connections are not intended to provide any mechanical support for the components or supporting wires.

Procedure

The solid and stranded wires used in this exercise will be prepared according to the terminal board wire list which follows. All of the stranded wires must be tinned before termination, but the solid wires and resistor leads do not have to be tinned.

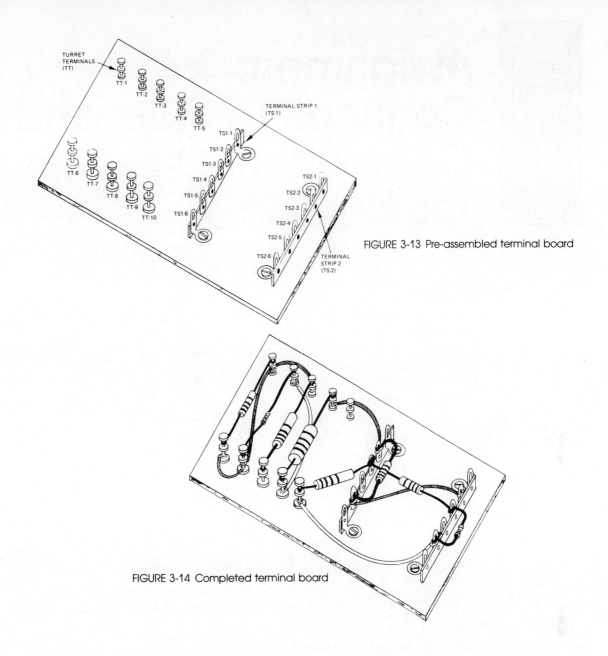

FIGURE 3-13 Pre-assembled terminal board

FIGURE 3-14 Completed terminal board

Examine the pre-assembled terminal board (figure 3-13). Notice the field of ten turret terminals (TTs) and the numbers assigned to each terminal (1–10). Also examine the six-lug terminal strips (TSs) and the numbers assigned to each terminal lug. The first digit of the hyphenated number represents the terminal strip and the second digit represents the terminal lug. For example, TS1-1 represents terminal strip 1, terminal lug 1 and TS2-5 represents terminal strip 2, terminal lug 5. Terminal number assignments are shown in figure 3-13.

Inspect all hand tools to be used in this exercise to insure that they are in safe working condition. Check for:

- Condition of the soldering iron tip
- A damp sponge in the soldering station
- Cutting edges on wire strippers and diagonal cutters
- Condition of the solder sucker

The completed terminal board with all required wires and resistors terminated according to the wire list is shown in figure 3-14. Refer to Table 3-2 for wire dimensions and destinations of all wires and resistors. It is suggested that all wires be terminated before resistor leads.

TABLE 3-2 Terminal Board Wire List

TERMINATION	WIRE TYPE	LENGTH	FROM	TO
1	Solid	2.0" (5.08 cm)	TT-1	TT-3
2	Solid	1.5" (3.81 cm)	TT-6	TT-7
3	Solid	2.5" (6.35 cm)	TT-1	TT-7
4	Stranded	2.5" (6.35 cm)	TT-2	TT-9
5	Stranded	3.0" (7.62 cm)	TT-4	TS1-2
6	Stranded	3.5" (8.89 cm)	TT-10	TS2-5
7	Solid	3.5" (8.89 cm)	TS1-5	TS2-2
8	½ W lead	Equidistant between terminal posts	TT-1	TT-6
9	¼ W lead	Equidistant between terminal posts	TT-2	TT-7
10	1 W lead	Equidistant between terminal posts	TT-3	TT-8
11	2 W lead	Equidistant between terminal posts	TT-4	TT-9
12	1 W lead	Equidistant between terminal posts	TT-10	TS1-3
13	½ W lead	Equidistant between terminal posts	TS1-2	TS1-5
14	½ W lead	Equidistant between terminal posts	TS1-3	TS2-3
15	¼ W lead	Equidistant between terminal posts	TS2-3	TS2-5

The illustrations in figure 3-15 are provided to aid in the terminating and crimping process.

NOTE: When connecting more than one wire on the same terminal, the final resting position for both wires should be close to each other. This will provide a more efficient and reliable means of connecting the wires during soldering.

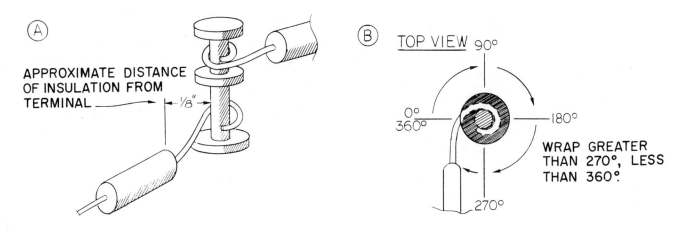

FIGURE 3-15 Wire crimping technique for turret terminals

94 Assignments

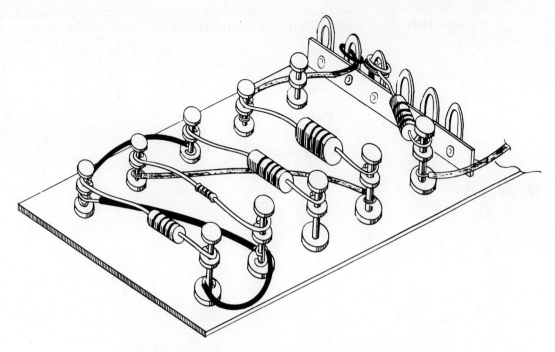

FIGURE 3-16 Close-up of terminal board around turret terminals

Following are the steps for crimping wire on terminal strip lugs:

1. Strip approximately 1" (2.54 cm) of the wire; tin stranded wire.

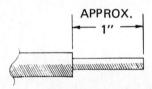

2. Feed the wire through the loop of the lug and bend it at a right angle to the lug.

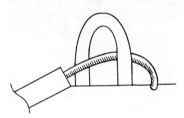

3. (A) Upper wrap—place the free end of the wire *over* the section leading into the lug.
   (B) Lower wrap—place the free end of the wire *under* the section leading into the lug.

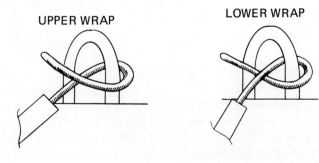

Assignment 3   95

4. Wrap the free end of the wire around the edge of the lug and snip the end with diagonal cutters so that it meets the section leading into the lug. The free end should not overlap the first loop (see figure 3-17).

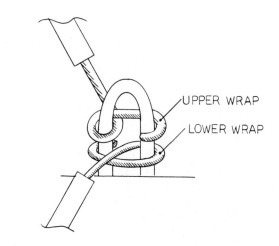

FIGURE 3-17 Crimp for two wires on the same terminal lug

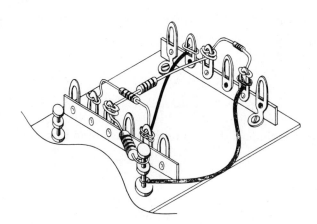

FIGURE 3-18 Close-up view of terminal board around terminal strips

A close-up view of the terminal board around the terminal strips is shown in figure 3-18. Inspect your work for the following crimping problems (see figure 3-19):

A. Loose crimps around the terminal posts of the turret terminals or around the edges of the terminal lugs.
B. Crooked crimps—wraps that are not parallel with the base of the terminal board.
C. Separated crimps—Terminals that contain more than one lead or wire may separate from one another. Individual crimps should be held tightly together.
D. Too much crimp—an excessive amount of loops. For turret terminals, the loops should be greater than 270 degrees but less than 360 degrees from the beginning of wrap. For terminal strip lugs, the loose end of the wrap should meet with the beginning of the wire leading into the lug.

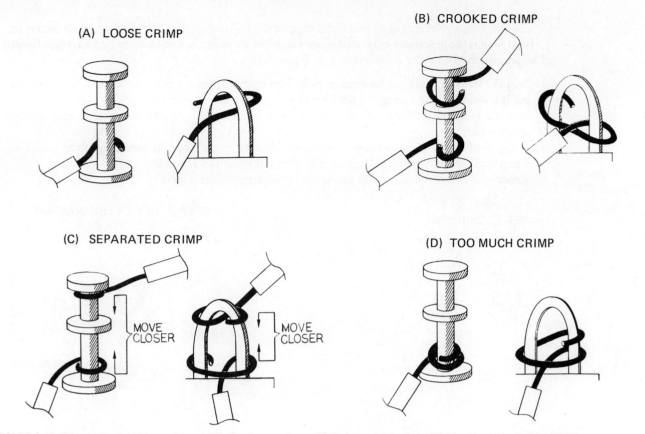

FIGURE 3-19 Common crimping problems (A) The loose crimp (B) The crooked crimp (C) The separated crimp (D) Too much crimp

EXERCISE 2: SOLDERING THE CRIMPED TERMINATIONS

The procedure involved in soldering the crimped terminals is similar to the procedure used to solder the wire splices. The molten solder should flow through and wet all of the wires around the terminal post. The solder feeding skill developed in Assignment 2 will help you to produce the proper wetting action and to control the quantity of solder applied to terminations.

Procedure

Fasten the terminal board to the PCB holder which provides a rigid support. Adjust the height and position of the PCB holder so that it is in a comfortable working position. Solder the terminations making sure that the molten solder covers all of the wires around the terminal post.

Inspect the soldered terminations for the following imperfections:

- Burnt insulation
- Solder wicking
- Excessive solder
- Cold-soldered joints
- Insufficient wetting action
- Crimp separation

EXERCISE 3: THE DESOLDERING PROCESS

Desoldering involves two basic steps: the defective component or wire must be removed without altering the condition of any of the adjacent wires or components and the affected area must be prepared for easy insertion of the new wire or component.

Procedure

Following is the procedure to be used to remove all of the wires and resistors from the terminal board and to prepare the board for a fresh start (see figure 3-20).

1. Snip the leads to break the continuous path. The length of the cut from the terminal is not critical, but it should be long enough to hold between your fingers or with your needle nose pliers.
2. Heat the terminal with the soldering iron and, at the same time, remove the excess solder with the solder sucker.
3. Using needle nose pliers, remove the lead by twisting in the opposite direction of the wrapping. Remove all of the wires and leads on the terminals using this method.
4. Reheat the terminal and remove excess solder with the solder sucker.

STEP 1: SNIP LEAD

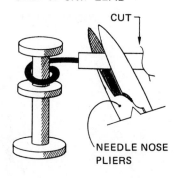

STEP 2: HEAT TERMINAL AND SUCK SOLDER

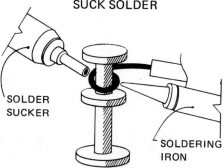

STEP 3: REMOVE LEAD

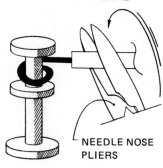

STEP 4: REMOVE EXCESS SOLDER

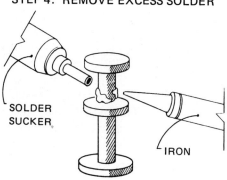

FIGURE 3-20 The desoldering removal procedure

QA CHECKLIST: ASSIGNMENT 3

Exercise 1: Wire Crimping and Component Mounting

Check for:

1. Improperly tinned stranded wire
2. Uniformity of wraps around turret terminals—more than 270 degrees, less than 360 degrees
3. Uniformity of wraps around terminal strips
4. Proper distance between insulation and joint
5. Accuracy of wires and resistors according to wire schedule

Exercise 2: Soldering the Crimped Terminations

Check for:

1. Solder wicking
2. Burnt insulations
3. Excessive solder coverage
4. Insufficient solder coverage
5. Cold-solder joints

Exercise 3: The Desoldering Process

Check for:

1. Damaged terminals
2. Insufficient removal of solder
3. Burnt or discolored terminal board due to excessive heat

Assignment 4: Coaxial and Shielded-Pair Cable Assembly

PURPOSE

- To become skillful in working with two types of shielded cable.
- To introduce general cable assembly techniques that can be used for shielded cable assembly.
- To interpret cable assembly procedures supplied by connector manufacturers.
- To test assembled cable using the ohmmeter.

COMPONENTS & EQUIPMENT

Components

- Approximately 40" (102 cm) of TSP (twisted-shielded-pair) cable
- Approximately 40" (102 cm) of RG-58 coaxial cable
- Four alligator clips with insulating boots (two red and two black)
- One BNC connector plug kit
- One dual-banana plug
- Two male banana plugs (black)
- Heat-shrink tubing—four pieces of 1" (2.54 cm) long, ⅜" (0.95 cm) diameter tubing; two pieces of 1.5" (3.81 cm) long, ½" diameter tubing.
- Approximately 10" (25.4 cm) of spaghetti insulation sleeving.

Equipment

- Basic hand tools
- PCB holder
- Soldering station and solder
- Metric or English scale
- Heat gun

EXERCISE 1: TSP (TWISTED-SHIELDED-PAIR) CABLE ASSEMBLY

The TSP cable, which will be assembled in this exercise, has one end connected to a standard dual-banana plug and a single banana plug and the opposite end connected to a pair of alligator clips, color-coded, insulated boots, and a single banana plug. The completed cable assembly is illustrated in figure 3-21.

Stripping both types of shielded cable was covered in Assignment 1. One obvious difference between the TSP and the coaxial cables is the shield structure that surrounds the inner conductor(s). Another

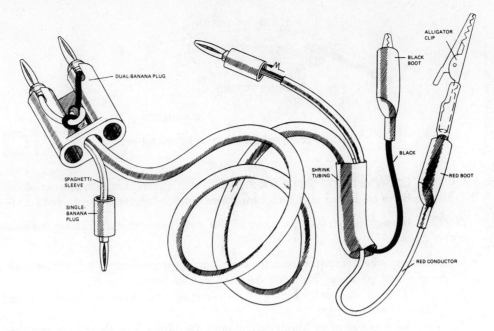

FIGURE 3-21 Completed TSP cable assembly

important difference is the electrical characteristics of these cables. The TSP does not have a characteristic electrical impedance while the coaxial cable does (RG-58 is a 50-ohm cable). The electrical nature of each cable, therefore, determines its principle application as an electrical conductive medium. Generally, the TSP cable is used on power supply output leads or low-frequency signal applications and the coaxial cable is used on high-frequency signal sources or as a direct (1:1) oscilloscope probe.

Procedure

Prepare the TSP cable for assembly by removing 5" (12.7 cm) of outer insulation from one end of this cable. This end will be terminated to the dual banana plug and the single banana plug. Remove approximately 6" (15.2 cm) of outer insulation from the opposite end. This end will be terminated to two alligator clips and a single banana plug. Inspect the cable for the following problems before proceeding to the next step:

- Damaged insulation to internal conductors.
- Damage to shield-conductor lead.

Initial Cable Preparation (see figure 3-22).

1. Remove foil (shield) from both ends of the TSP cable *up to the edge of the outer gray insulation*.
2. Strip approximately 0.5" ± 0.125" (1.27 cm) of insulation from both ends of each pair of inner conductors. Tin all stripped leads and be especially careful that solder wicking does not occur.

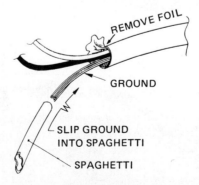

FIGURE 3-22 Initial cable preparation

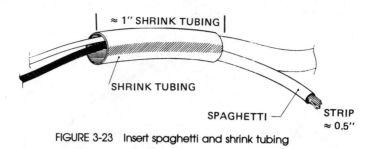

FIGURE 3-23 Insert spaghetti and shrink tubing

Insertion of Spaghetti and Shrink Tubing. This step insures that each inner-insulated conductor and the shield lead are reinforced at the junction where outer insulation was removed (see figure 3-23).

NOTE: Although the ends of this cable were stripped to different lengths, the following instructions apply to both ends of the cable.

1. Slide the spaghetti insulation over the cable, leaving approximately ½" of bare wire extending from the spaghetti insulation. Bend this insulated lead as illustrated in figure 3-23.
2. Slide a 1" piece of shrink tubing (⅜" diameter) over the inner conductor pair and the spaghetti-insulated shield lead.
3. Apply the heat gun to the shrink tubing until the tubing is firmly shrunk over the junction point.

CAUTION: Extensive application of heat on shrink-tubing material will not cause additional shrinkage—it will burn adjacent wire insulation. The degree of shrinkage, which is controlled by the manufacturer, is normally a specific percentage of its original dimensions.

Lead Preparation. Instructions are given for each end of the cable to insure proper preparation of the leads, which are terminated to different connectors—the alligator clips, the single banana plug, and the dual banana plug.

1. The alligator clips and single banana plug will first be terminated to the end where 5" (12.7 cm) of outer insulation was removed. First connect the red and black inner conductors to the alligator clips. Then terminate the spaghetti-insulated shielded lead to the single (black) banana plug. Follow the illustrations in figure 3-24.

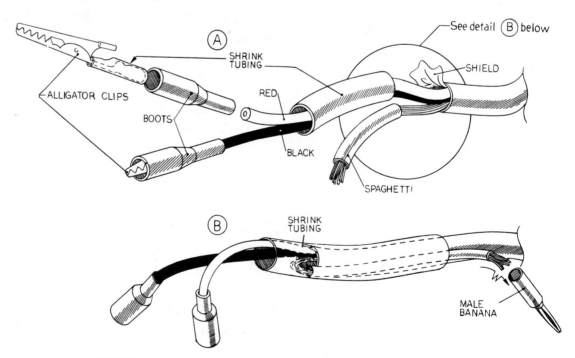

FIGURE 3-24 End with alligator clip and single banana plug

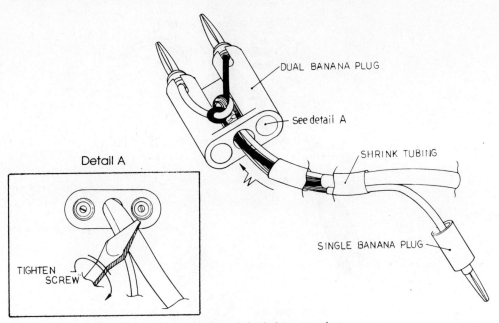

FIGURE 3-25 End with dual and single banana plug

2. The dual and single banana plugs are terminated to the end where 6" (15.2 cm) of outer insulation was removed. Connect the inner red and black conductors to the standard dual banana plug. Both the red and black inner conductors must be cut to fit to accommodate the terminal screws on the dual banana plug (see detail A of figure 3-25). Terminate the spaghetti-insulated shielded lead to the single banana plug. Follow the illustrations in figure 3-25 to complete this end of the cable. Your completed TSP cable will be tested later in Exercise 3.

EXERCISE 2: COAXIAL CABLE ASSEMBLY

Unlike the TSP cable assembly procedure, the coaxial cable assembly procedure requires more accurate measurement and stripping techniques. A careful inspection of this type of shielded cable shows an inner conductor surrounded by a braided structure that extends the entire length of the cable. The diameter of the inner insulation combined with the diameter of the inner conductor and its surrounding braided structure causes the electrical characteristic called impedance. The inner insulating material is also referred to as a dielectric.

As in the TSP cable assembly procedure, the preliminary cable preparation is the critical stage for the coaxial cable procedure assembly. The completed coaxial cable assembly is illustrated in figure 3-26.

Procedure

Preparation of the Alligator Clip End. The following instructions apply *only* to the end of the coax where the greater length of outer insulation was removed (6", or 15.2 cm). This end of the cable will be connected to alligator clips insulated with rubber boots.

The diagrams in figure 3-27 illustrate the correct procedure for fishing out the insulated center conductor through the braided shield structure.

   A. Insert a pointed tool (scribe) into the braided area approximately 0.25" (0.635 cm) from the edge of the outer insulator.
   B. Separate the braided structure with the point of the tool to expose at least the width of the inner dielectric material.
   C. Insert the pointed end of the tool into the open braided structure and under the inner dielectric.
   D. Bend the cable at the point where the tool was inserted with the tool under the dielectric as a pivot.

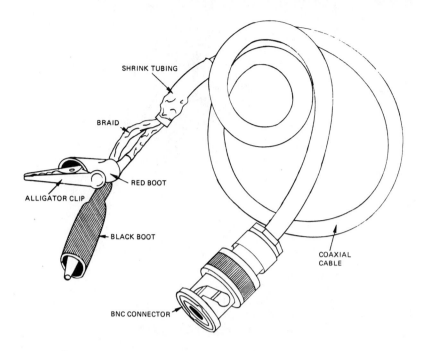

FIGURE 3-26 Completed coaxial cable

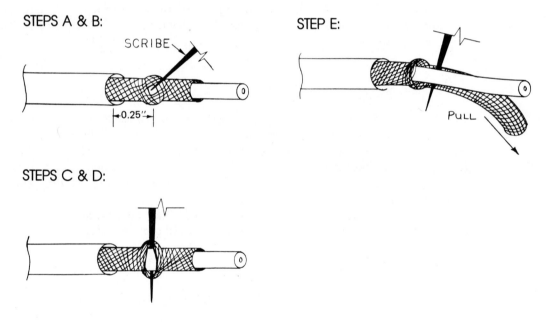

FIGURE 3-27 Coax preparation—alligator clip end

E. Slowly, pull the inner conductor completely out through the opening in the braided structure with the tool as the main leverage instrument.
F. Pull gently on the end of the braid and patch the operating area by dressing any loose strands from the braided structure.
G. Slide approximately 1.5" (3.81 cm) of ½" diameter shrink tubing over the junction of the shielded braid and the inner-insulated conductor. Apply heat to the shrink tubing with a heat gun as in the TSP cable preparation.
H. Strip approximately 0.5" ± 0.125" (1.27 cm) of insulation from the inner conductor.

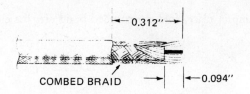

FIGURE 3-28 Cable preparation for BNC end

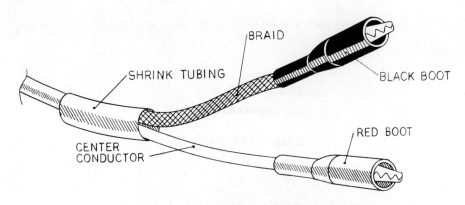

FIGURE 3-29 Coax end to alligator clips

FIGURE 3-30 BNC connector nomenclature

Preparation of the BNC End. Using the dimensions shown in figure 3-28, prepare the BNC end of the cable. Fray the "combed" braid with the pointed tool. Strip 0.094" ± .005" (0.238 cm) from the inner insulation.

The coaxial cable is now ready to be connected to the alligator clips on one end and to the BNC connector on the opposite end.

Connection of the Alligator Clip End. The inner conductor is terminated to the alligator clip with the red insulating boot and the braid is terminated to the alligator clip with the black insulating boot. The connections of the inner conductor and the braid to their respective alligator clips are shown in figure 3-29.

Connection of the BNC Connector End. Following are the steps for connecting the BNC connector to the cable (see figure 3-30 for illustrations of the devices used in this procedure):

1. Slide a nut, washer, gasket, and clamp over the braid. Slide the clamp in position so that it fits squarely against the end of the outer insulation (the cable jacket).

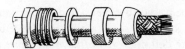

Assignment 4   105

2. With the clamp in place, fold the combed braid *over* the clamp and trim it to 3/32" (0.238 cm).

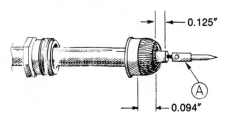

3. Slip the contact in place and solder it as illustrated in detail A.

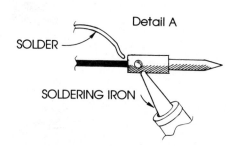

4. Push the assembly into the body as far as it will go. Slide the nut into the body and tighten it with a wrench. When tightening, hold the shell rigid and rotate the nut.

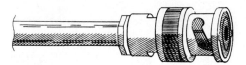

EXERCISE 3: CABLE INSPECTION AND OHMMETER TESTING

Manufacturers of pre-assembled cable are very concerned with the quality control of all cable assemblies for specific user functions. Cable users look at electrical characteristics (i.e., impedance, conductor resistance, and insulation resistance) and mechanical characteristics of cables when choosing pre-assembled cable. Consequently, cable manufacturers must provide electrical and mechanical specifications collected from actual laboratory tests.

In this exercise, the cables assembled in Exercises 1 and 2 will be tested in two categories. In the first test, the electrical continuity between the ends of the cable will be determined and in the second test, mechanical stress will be placed on the cable assembly. The principal testing instrument for both tests is the ohmmeter.

Procedure

Continuity Test. Following are the steps for performing the continuity test. This method can also be used to find cable shorts.

1. Place the ohmmeter range switch on the lowest position—R × 1 ohm. Calibrate both ends of the scale with the zero and ohm adjustments.
2. Clip one of the ohmmeter test leads to one end of your cable and the other test lead to the opposite cable end. If the initial indication is 0 ohms, rotate the range switch to the next highest range. Do this until you reach the highest range.
3. An indication other than 0 ohms may suggest three conditions: a high resistance connection, an open connection, or an incorrect connection.
4. Your cable fault can be traced to either a bad solder connection (a cold-solder joint) or a no-solder joint. The cold-solder joint will show a high resistance in the higher range. A no-solder connection will indicate an infinite deflection *regardless of the range position* on the ohmmeter.

Mechanical (Pull) Test. The strength of the cable assembly is determined in the mechanical (pull) test. Cables that are exposed to the repetitive actions of insertion and removal from instrument input or output connectors are stressed at specific areas of the cable. The major stress is normally placed on the cable connector ends. Following are the steps for performing the mechanical test:

1. Connect the ohmmeter as in the continuity test.
2. Apply a gentle twisting and pulling action to the cable and note the ohmmeter indications as these external mechanical stresses are applied. Intermittent indications signal a poor electrical connection which should be corrected.

QA CHECKLIST: ASSIGNMENT 4

NOTE: As it is extremely difficult to examine intermediate steps once the cable is totally assembled, it is suggested that each stage of the process be examined before proceeding to the next.

Exercise 1: TSP Cable Assembly

Check for:

1. Proper insulation removal
2. Proper tinning of stranded wires
3. Proper application of shrink tubing
4. Proper termination of alligator clips and banana plugs

Exercise 2: Coaxial Cable Assembly

Check for:

1. Proper removal of insulation and correct braid preparation
2. Correct preparation of BNC connector end from given dimensions
3. Proper termination of alligator clips
4. Correct sequence of BNC connector sections—nut, washer, gasket, clamp, contact, and body
5. Proper braid preparation over BNC connector clamp
6. Proper soldering of BNC connector contact

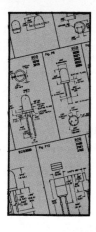

Assignment 5: Printed-Circuit Board Assembly Techniques

PURPOSE

- To develop the technique of mounting radial-lead and axial-lead components.
- To develop skill in soldering single-sided and two-sided PCBs.
- To develop desoldering techniques on single-sided and two-sided PCBs.
- To develop techniques of repairing and replacing existing components on PCBs.

COMPONENTS & EQUIPMENT

Components

- One single-sided PCB and one two-sided PCB
- Component kit (assortment of resistors, capacitors, diodes, transistors and IC DIP Package)
- Pre-assembled PCB

Equipment

- Soldering station and solder
- Basic hand tools
- PCB holder
- Miscellaneous equipment (solder wick, lacquer thinner)

EXERCISE 1: COMPONENT IDENTIFICATION AND MOUNTING

Radial-lead and axial-lead components will be used for this exercise. Radial-lead components can be identified by the leads extending from the *same side or end* of the main component body, while axial-lead components can be identified by the leads extending from *opposite ends* of the component body (see figure 3-31).

Passive components such as resistors, capacitors, and inductors (chokes) may be purchased in the radial-lead or axial-lead configuration. The selection of the lead configuration is normally determined by availability or by spacial considerations.

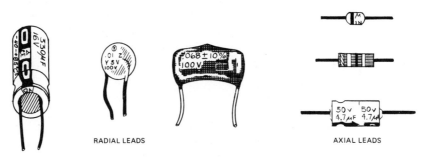

FIGURE 3-31 Radial vs axial lead

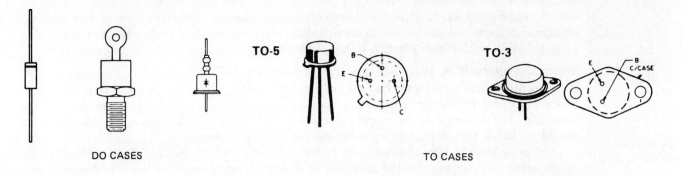

DO CASES   TO CASES

FIGURE 3-32 Common diode and transistor case styles

FIGURE 3-33 Integrated circuit package styles: (A) TO-5 style, (B) 14-pin dual-in-line package (DIP)

Active components such as diodes, transistors, and integrated circuits are not normally identified by the radial-lead or axial-lead configuration. Instead, a standardized package identification system adopted by an organization known as the Joint Electronic Devices Engineering Council (JEDEC) is used by many electronic manufacturing companies. For example, semiconductor diodes are identified by a *DO* (diode outline) number such as DO-7 or DO-16. Transistors and integrated circuits (ICs) with similar case styles are identified by a *TO* (transistor outline) number such as TO-18, TO-5, or TO-3 (see figure 3-32).

Most integrated circuits are packaged in one of the package styles shown in figure 3-33. The TO style consists of a header with sealed leads, on which the transistor or device is mounted and hermetically sealed. The number of leads varies from three to twelve or more. The dual-in-line (DIP) package may be made of ceramic or plastic material and may contain eight, twelve, sixteen, or more leads. An even larger size IC package is used to house large-scale-integration (LSI) assemblies which may have as many as several hundred circuits on a single chip. The leads are brought out on two or more sides of the package.

Index markings that identify component elements as the base emitter, and collector for a bipolar transistor can be found on the specification sheet for the specific device. For example, the index tab included on a TO-22 case for a 2N2222 bipolar transistor identifies the location of the emitter element.

Procedure

Remove all of the components from the component kit and determine the physical style (radial, axial, DO, TO, flat-pack, or DIP case) for each of the components. It is not necessary to determine the exact *value*; determination of the style is more important.

There are several methods of preparing (forming) component and device leads. The method used depends on the method selected to solder the printed-circuit boards. Soldering methods such as resistance soldering and dip and wave soldering are commonly used in industrial production. Because these methods are somewhat specialized and are used in mass-production applications, they are beyond the scope of this textbook, which will focus on *prototype*, or one-of-a-kind fabrication using hand-soldering techniques.

Preparation of the PCB. When bare copper PCB is left exposed to the atmosphere, a microscopic film of copper oxide forms on the surface of the foil. Unless this oxide is removed, it will prevent the solder from alloying properly with the copper during the soldering process. This will result in a defective soldered connection. The formation of the copper oxide film is very rapid. For this reason, the foil should be cleaned just before the components are assembled and leads are soldered.

Following is one method of cleaning the PCB for prototype construction (see figure 3-34). Wet the copper surface with tap water and scrub it with an abrasive cleaner (pumice) or wet-dry, abrasive paper (extra fine grit). After scrubbing, rinse the board thoroughly to remove all abrasive residue and dry it with a paper towel. When the PCB is dry, apply a very thin coating of diluted soldering flux to the copper foil and allow it to air dry. The PCB is now ready for the assembly procedure.

Examine the single-sided (copper foil on only one side) and the double-sided (foil on both sides) PCBs. What is the difference between these PCBs besides the single-sided and double-sided features? The single-sided PCB has a dull appearance because it is oxidized copper foil, while the double-sided PCB has a shiny appearance because it is plated with a lead-tin alloy. In manufacturing, the double-sided PCB was subjected to an additional step after the etching process. This step protects the copper foil from atmospheric oxidation; therefore, this board is ready for assembly—it does not have to be cleaned. The single-sided board, however, is contaminated with copper oxides and must be cleaned by the preceding procedure.

NOTE: *Cleaning the bare copper foil areas just prior to soldering is essential and cannot be emphasized enough. Poor soldering will result if the copper foil is contaminated by any type of oxidation.*

Preparation of Components for Mounting on the PCB. In general, all component bodies should be mounted flush to the board if possible. This is particularly true of components that do not generate an excessive amount of heat. Following are guidelines for the preparation of component leads prior to insertion into the PCB.

To form axial-lead components, bend the lead at a 90-degree angle to the body with a fine needle nose plier (see figure 3-35). Grip the lead with the tip of the pliers at the point at which the bend is to be formed. Then bend the lead down firmly against the side of the plier nose to form the 90-degree bend.

The component lead should extend from the body of the component a minimum distance equal to the diameter of the lead before the start of the bend (see figure 3-36).

NOTE: *"L" must be equal to or greater than "D."*

The bend radius must be no less than ½ the lead diameter (see figure 3-37).

NOTE: *"R" must be greater than ½ "D."*

Ginked or beaded radial components must be no higher than 0.1" (0.254 cm) over adjacent components with similar leads. Ginked components have leads formed in such a manner to allow components to stand off the PCB (see figure 3-38).

The bottoms of straight-lead (radial) components—non-ginked and non-beaded—must be no more than 0.2" (0.5 cm) off the PCB surface (see figure 3-39).

The following conditions are unacceptable in regard to the physical damage of components:

1. Any damage that electrically degrades the component.
2. Chipping or cracking of hermetic seals.
3. Distortion of components or wire insulation, which could affect performance, serviceability, reliability, or safety.
4. Rusted, corroded, or water-damaged components.
5. Axial components that have damaged or cracked conformal coating inside the outer edge of the caps (see figure 3-40).

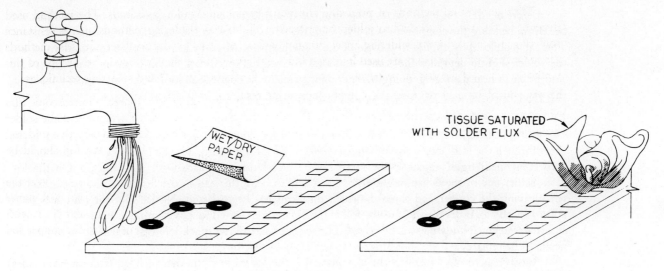

FIGURE 3-34 PCB cleaning process

FIGURE 3-35 Forming axial lead components

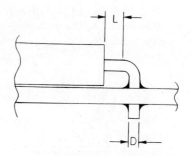

FIGURE 3-36 Lead bend

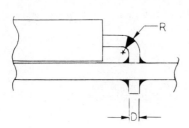

FIGURE 3-37 Bend radius

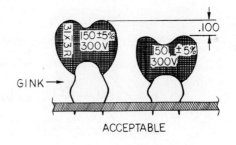

FIGURE 3-38 Ginked component height limitations

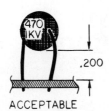

FIGURE 3-39 Radial component lead limitations

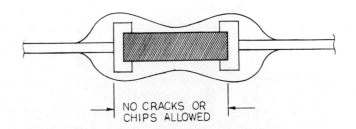

FIGURE 3-40 Damaged conformal coating

Assignment 5   111

Before bending the leads that will be terminated in soldered connections, one must be certain that the benefit in bending the component will outweigh the stress applied to the leads and the component body. This same philosophy applies to the use of the soldering iron. Anytime heat is applied to a soldered joint, the solder slightly degrades the component's rated electrical specifications. Since many components are susceptible to heat damage, redoing a solder joint to make the connection look better may cause a failure in the component or may degrade the component's value resulting in possible failure at a later point in time.

Mounting Leads on the PCB. When all of the axial-lead and radial-lead components are formed, insert them through the lead access holes *from the insulated side* of the board, not the foil side. Press the body of each component tightly against the board surface and hold it in place with your fingers. On the foil (conductor) side of the board, use needle nose pliers to bend the lead in the direction of any conductive path leaving the copper pad through which that lead has been passed. Notice that the lead is bent at approximately a 30-degree angle to the foil in the direction of the conductive path (see figure 3-41). This is a partial lead cinch known as a *service bend*. The service bend allows quick removal of leads if a component needs to be replaced.

Another method of forming the termination is the *full-cinch* method, which is a more mechanically secure component mounting technique than the service bend. The full-cinch bend is made by pressing the flat blade surface of a screwdriver firmly against the lead onto the terminal pad (see figure 3-42). When

(A) Lead inserted from the component side of board

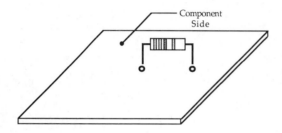

(B) Lead bent with the aid of needle nose pliers approximately 30 degrees in the direction of the conductive path.

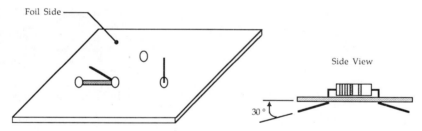

FIGURE 3-41 Technique of inserting leads in PCBs: (A) lead inserted from the insulated side of the board, (B) lead bent with the aid of needle nose pliers approximately 30 degrees in the direction of the conductive path

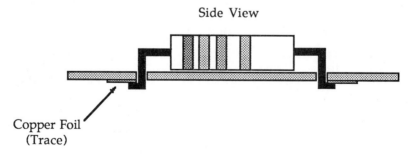

FIGURE 3-42 The full-cinch termination

this type of termination is used, the component is not easily removed from the PCB. The full-cinch mounting technique is used on printed-circuit boards that are exposed to extreme environmental conditions.

NOTE: Although both methods of bending components on the foil side have been discussed, it is suggested that the service bend be performed on the majority of your leads; however, you may experiment with the full-cinch termination if you wish to do so.

Soldering the PCB. After all of the components are mounted, the PCB is ready to be soldered. The soldering skills developed in the preceding soldering assignments will be used when soldering printed-circuit boards. A brief review of the important soldering techniques may be helpful at this point.

1. The selection of the proper soldering iron wattage rating and tip style is extremely important. *Never use a soldering gun* on a printed-circuit board. A 25-watt through 60-watt soldering iron with a screwdriver blade or a chisel tip is most commonly used. The tip should be in the temperature range of 600 to 650 degrees F (315 to 343 degrees C).
2. The tip should be cleaned periodically by wiping it across a damp sponge.
3. A small amount of solder should be applied to one side of the flat face of the tip to insure efficient heat transfer from the tip to the connection (solder bridge effect).
4. With the soldering iron held in a stationary position on the connection, a small amount of solder is applied to the *opposite* side of the lead and terminal pad. Solder is always applied to the *junction* of the lead and terminal pad area, *never to the tip* (see figure 3-43).

Solder all of the leads mounted on this single-sided PCB. Using diagonal cutters, snip the leads at the crest or peak of the soldered lead (see figure 3-44).

Cleaning the Soldered PCB. As was mentioned earlier, some electronic manufacturers find it undesirable to remove the rosin flux from the PCB after the soldering operation; however, for our purposes, all soldered PCBs will be cleaned to provide a better surface for visible inspection.

The cleaning procedure requires the use of a cleaning agent and an acid brush to remove the rosin residue. The cleaning agent used here is lacquer thinner. The procedure is simple, but it requires some caution due to the flammable nature of the cleaning agent.

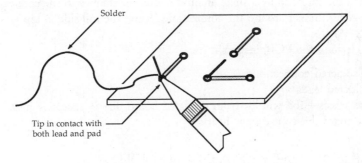

FIGURE 3-43   Correct PCB soldering process

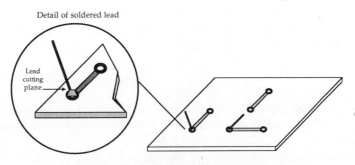

FIGURE 3-44   Cutting soldered leads

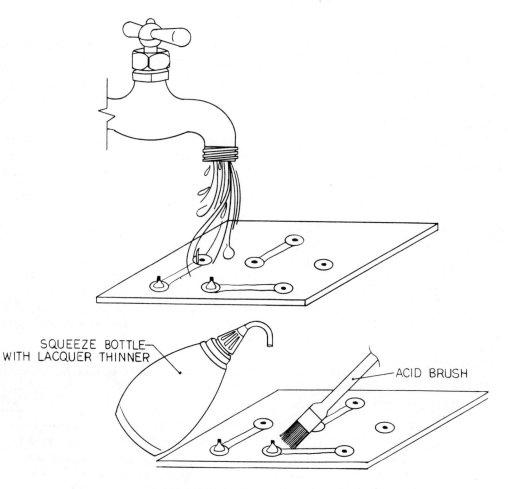

FIGURE 3-45 PCB cleaning procedure

To clean the PCB, apply a small amount of lacquer thinner to the connection. Using the acid brush, work the lacquer thinner around the connections. Rinse the foil side in tap water and dry it with a towel (see figure 3-45).

After cleaning the PCB, inspect it for:

1. Cold-soldered connections
2. Unsoldered leads
3. Raised traces—foil will lift from insulated board due to excessive heat
4. Burnt board due to excessive heat

EXERCISE 2: ASSEMBLY OF A DOUBLE-SIDED PCB

The double-sided PCBs used in this exercise are edge-connector, card-type boards designed to be inserted into an instrument's mainframe. A careful inspection of a sample board reveals that the holes provided for lead insertion are *plate-thru* holes; that is, the pads on each side of the board are connected through the hole by a conductive layer which is attached to the wall of the hole. This provides the required conductive path to opposite sides of the board.

Procedure

NOTE: The double-sided PCB on which you are working may be slightly different from the board illustrated in figure 3-46. For this reason, the locations given in this exercise for mounting components may not be the same on your board.

Select any ten components from your component kit. Your choices should include axial, radial, TO, and DIP packages.

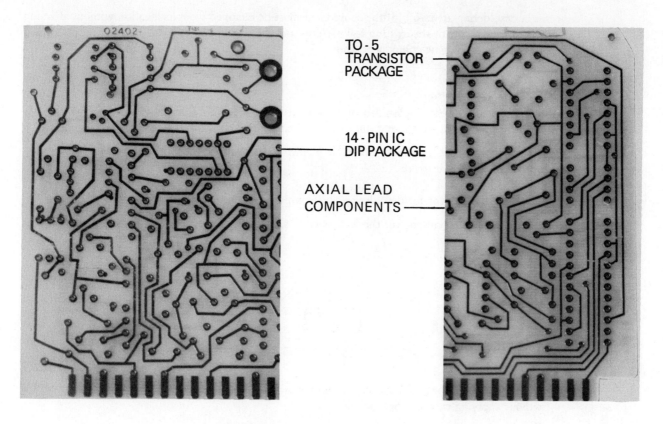

FIGURE 3-46 Recognition of specific hole patterns

An inspection of the sample PCB in figure 3-46 reveals a definite pattern of holes; that is, the matched pairs of holes are normally in a horizontal or vertical plane. The axial lead components, therefore, should be mounted in either the horizontal or vertical direction, *never* in a diagonal or angular plane with the edges of the board.

To find the transistor hole pattern look for a triangular pattern on the sample board.

Integrated circuit (IC) patterns may be easily recognized by the horizontal row of 4, 7, or 8 holes. The DIP package should be inserted at this pattern. For the TO style IC package, a circular pattern with eight or ten holes is common.

Mount and solder the ten components to the hole patterns on the double-sided PCB.

EXERCISE 3: COMPONENT REPLACEMENT AND REPAIR OF THE ASSEMBLED PCB

Once a lead from a component has been soldered to a PCB pad, it may be a difficult task to desolder the lead and remove it without damaging the component or the pad on the circuit board. Several tools and techniques are available to aid in the desoldering process. Tools such as the solder sucker, the desoldering bulb, and the solder wick remove solder from the pad and the component lead when the solder is in its liquid state.

Semiconductor devices, which are very sensitive to heat, can be destroyed if excessive heat is applied during the desoldering operation. In addition, the adhesive bond between the copper pad and the insulating material can be destroyed resulting in pad lifting. It is important, therefore, to choose a desoldering method with which the possibility of component or board damage is minimal. Because of its inherent heat-sinking ability, the solder wick is preferred over the solder sucker.

Although there are many acceptable methods of desoldering any component on the assembled PCB, only one method will be presented in this exercise. Regardless of the technique employed, the primary concerns in desoldering are twofold: the component must be removed from its location without affecting the adjacent components or the existing foil patterns and the vacant pad terminal must then be prepared for easy insertion of the new component leads.

Procedure

Using a pre-assembled, single-sided or two-sided board, identify some of the components on the board. The first five components that must be removed from the PCB are the axial lead components.

NOTE: Although a new component will not be inserted in the vacant spot created, the vacant pads must be prepared to accept a new component.

Following are the steps for desoldering and component removal:

1. Using the diagonal cutters, cut the lead as close as possible to the body of the component.

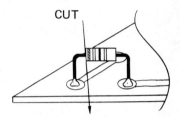

2. Use a solder wick (or solder sucker) to remove as much solder around the snipped leads as possible without applying excessive heat to the pad.

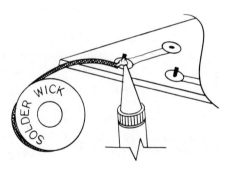

3. Remove the lead with the needle nose pliers. If this is not easily done, heat the lead (not the pad) with the soldering iron while simultaneously pulling the lead with the needle nose pliers.

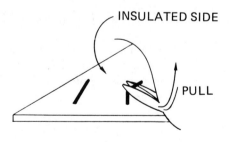

4. Using a solder wick or a solder sucker, prepare the vacant pad for easy insertion of the new component.

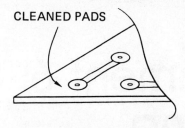

This desoldering method can be applied to any component on the PCB. The initial step of cutting the component leads is performed to provide a lead to grip with the needle nose pliers. If a lead is not easily accessible, the solder must be removed with the use of solder wick or solder sucker as the first step. The component can then be pryed or lifted with a soldering aid as the pads are reheated.

Finally, remove the following components from the sample board: the five axial components, the five radial components, the two TO-package components, and the two DIP integrated-circuit packages.

QA CHECKLIST: ASSIGNMENT 5

Exercise 1: Component Identification and Mounting

Check for:

1. Proper formation of radial-lead and axial-lead components
2. Proper insertion of components into pads
3. Adequate solder coverage on leads and pads
4. Excess component leads cut at the proper location
5. Proper removal of flux from soldered pads

Exercise 2: Assembly of a Double-sided PCB

Check for:

1. Proper lead preparation and insertion into pads
2. Adequate coverage of solder on both sides of plate-thru holes
3. Proper removal of flux from both sides of the PCB

Exercise 3: Component Replacement and Repair of the Assembled PCB

Check for:

1. Proper removal of affected component
2. Proper removal of solder around affected pads and traces for easy insertion of new component
3. Damage to adjacent component or components
4. Unintentional removal of solder from adjacent pads

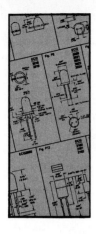

Assignment 6: Schematic Reading

PURPOSE
- To identify component symbols and their corresponding reference designations from actual schematic diagrams.
- To develop a schematic diagram from a component layout diagram of a basic RCL (resistor-capacitor-inductor) network or a simple electronic circuit.
- To develop a schematic diagram from an actual printed-circuit board.

EXERCISE 1: IDENTIFYING GRAPHIC SYMBOLS

The ability to recognize the symbol for any electronic component, connecting device, logic symbol, or other device on a schematic diagram requires a reliable reference guide. In Section 2 of this text, component symbols commonly found on schematic diagrams for consumer and industrial equipment are presented. More importantly, this large section covers the evolution and the philosophy of schematic symbology. It also provides illustrations of the actual components represented by each symbol group. The large variety of component styles makes it virtually impossible to illustrate even a small sample of each component group; however, this section will serve as our primary reference to this phase of reading schematic diagrams.

Procedure

The schematic diagrams to be used in this exercise are located in Section 2. They have been obtained from a number of sources, including industrial and commercial test equipment, consumer entertainment equipment, and electronic hobby magazines. There are twenty-five symbols to be identified in this exercise.

Identify the symbols on Worksheet 3-1, *Identification of Common Schematic Symbols*. Complete this worksheet in the following manner:

1. In the *Graphic Symbol* column, draw the symbol provided to you by your instructor. Be as specific as necessary when identifying each symbol. For example, identifying a transistor symbol as a transistor may not be adequate since there are several types of transistors each of which has its characteristic symbol modifier.
2. Complete the *Reference Designation* column with the appropriate letter designation for each component.
3. In the *Additional Information* column, note any information that is helpful to emphasize an important characteristic of the circuit or component.

WORKSHEET 3-1: Identification of Common Schematic Symbols

SYMBOL NUMBER	GRAPHIC SYMBOL	REFERENCE DESIGNATION	ADDITIONAL INFORMATION
1			
2			
3			
4			
5			
6			
7			
8			
9			

WORKSHEET 3-1 (continued)

SYMBOL NUMBER	GRAPHIC SYMBOL	REFERENCE DESIGNATION	ADDITIONAL INFORMATION
10			
11			
12			
13			
14			
15			
16			
17			

WORKSHEET 3-1 (continued)

SYMBOL NUMBER	GRAPHIC SYMBOL	REFERENCE DESIGNATION	ADDITIONAL INFORMATION
18			
19			
20			
21			
22			
23			
24			
25			

EXERCISE 2: DEVELOPING SCHEMATIC DIAGRAMS

The field of digital electronics has revolutionized many industrial processes in regard to the speed, accuracy, and quantity of complex calculations or operations performed. In the field of electronic drafting the computer is utilized as an important tool to eliminate tedious operations and to enhance complex processes such as multilayer, printed-circuit board design. Despite the influence of computer-assisted drafting on modern electronic graphic practices, however, the art of producing schematic diagrams from simple engineering sketches is still practiced. In this exercise, the skill of developing simple circuit diagrams from illustrations of assembled circuit networks on terminal strips and from PC boards will be developed.

Procedure

The following illustrations represent the three most common methods of terminating components and wires to rigid or semi-rigid mediums. Examine each illustration to determine the field of the circuit configuration on each terminal board; that is, the orientation of any passive component (resistor, capacitor, or inductor) in relation to any active component (diode, transistor, vacuum tube, or integrated circuit).

Develop the circuit diagram in the *standard position* as much as possible. A diagram drawn in the standard position has the circuit *inputs* drawn from the *left* side of the page and the *outputs* drawn from the *right* side. The terms inputs and outputs are used to describe signals or applied dc voltages to the circuit network. If both signals and dc voltages are present on a diagram, the standard position dictates that the input signals enter from the left side of the page and the output signals exit at the right side of the page. All dc voltages applied to the same circuit are placed in an area of the diagram where it will not be confusing to indicate a signal input or output point. The circuit diagram shown in figure 3-47 illustrates the standard position.

Circuit with Applied DC Voltage Only

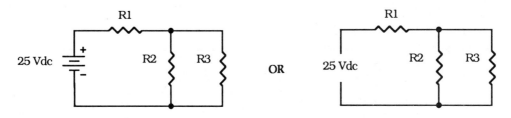

Circuit with Both Signal and Applied DC Voltage

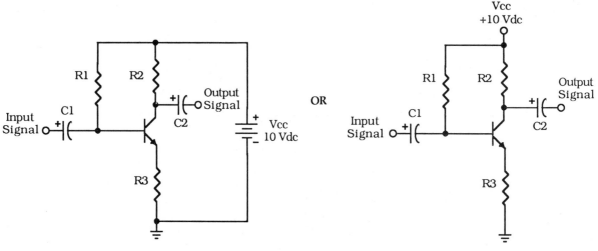

FIGURE 3-47 Circuit diagrams in standard position

WORKSHEET 3-2: Components Mounted Between Turret Terminals

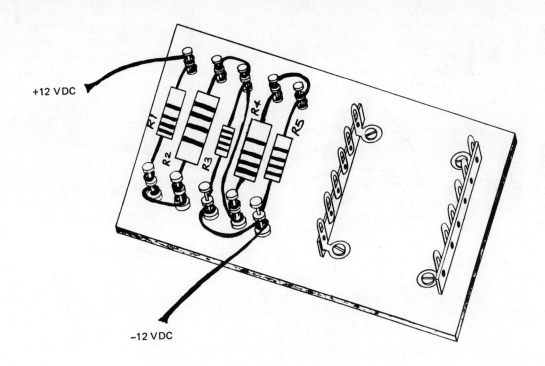

Circuit diagram of preceding network

WORKSHEET 3-3: Components Mounted on Terminal Strips

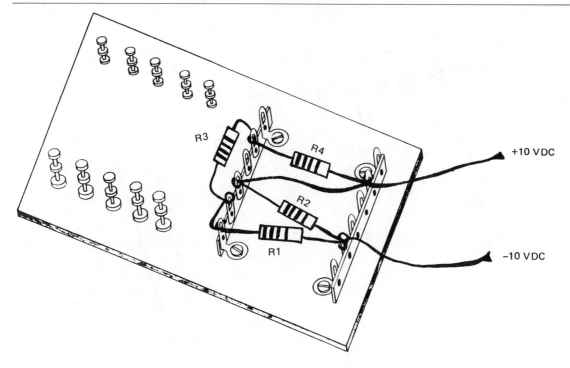

Circuit diagram of preceding network

WORKSHEET 3-4: Components Mounted on Printed-Circuit Boards

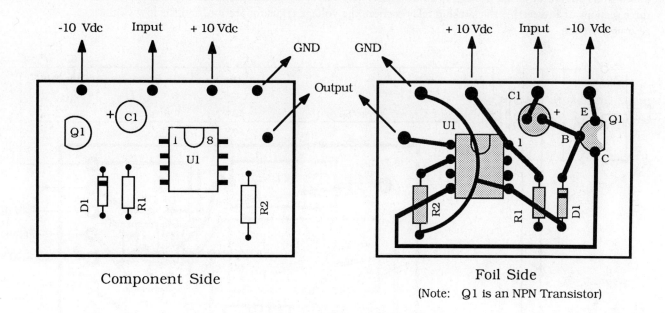

Component Side

Foil Side
(Note: Q1 is an NPN Transistor)

Circuit diagram of preceding network

Assignment 6   125

WORKSHEET 3-5: Circuit Diagram of Rectifier Board

The *rectifier board* is a subassembly that will be used in the following projects. The function of this board is to convert the ac line voltage to a dc voltage. The output dc voltage from this board is applied to three sections of Project 1—the latching relay section, the voltage regulator section, and the low voltage ac source section.

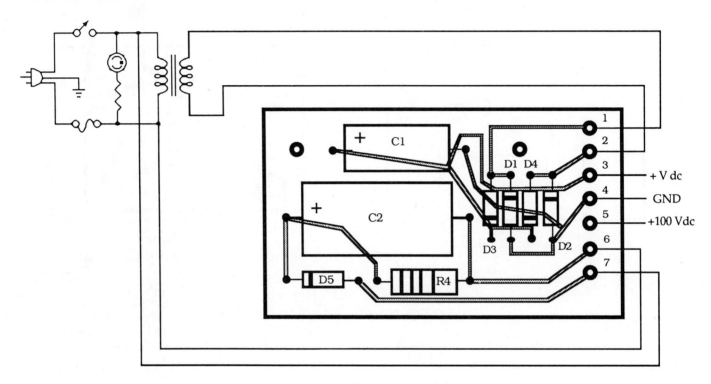

Partial Diagram of the main test fixture

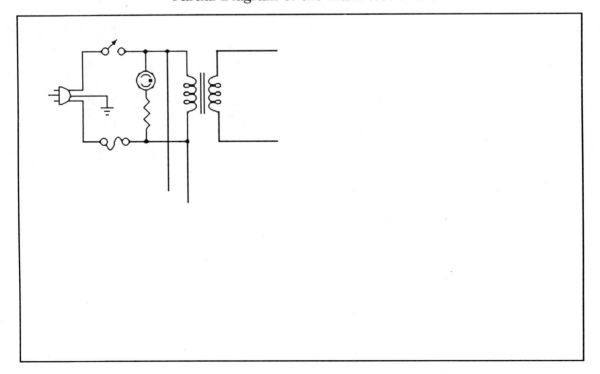

126   Assignments

QA CHECKLIST: ASSIGNMENT 6

Exercise 1: Identifying Graphic Symbols

Check for:

1. Accuracy in duplicating symbols
2. Correct reference designations
3. Inclusion of essential notations

Exercise 2: Developing Schematic Diagrams

Check for:

1. Accuracy in reproducing circuit networks
2. Drafting technique—component placement and standard-position requirement

SECTION 4
PROTOTYPE CONSTRUCTION PROJECTS

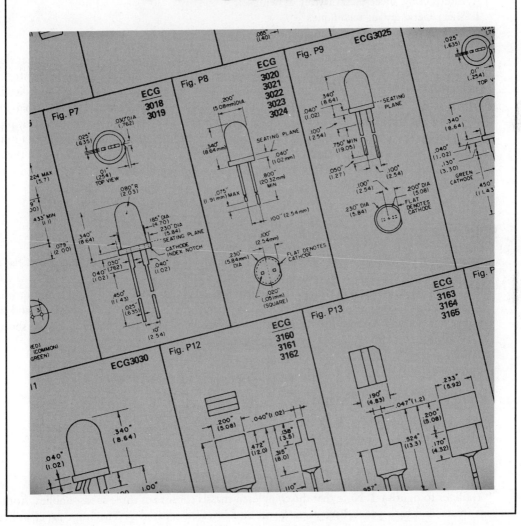

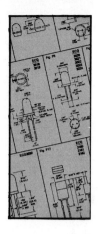

Project 1: Chassis Fabrication

The purpose of this assignment is to fabricate the chassis for Project 2. This procedure requires three basic chassis fabrication skills:

1. Interpreting sheet metal layout drawings
2. Sheet metal layout techniques
3. Using basic sheet metal equipment

Once the chassis layout drawings are read and interpreted properly, all necessary dimensions are then transferred to the flat sheet metal stock. Chassis fabrication begins with shearing the flat stock to the final perimeter dimensions. The following operation includes the production of all round holes by punching and/or drilling. This operation requires a clearly defined hole assignment schedule that identifies the location, size, and equipment to produce the required holes. The shaping operation employing sheet metal bending equipment is the final procedure for producing the finished product. Finishing or applying paint or other protective nonmetallic coating to the bare metal chassis is optional and will be left to the discretion of each individual. Since Project 1 is a prototype, the finishing requirement is not a mandatory procedure in the fabrication of this chassis.

PURPOSE

- Given a set of chassis layout drawings, the student will determine the style and size of flat sheet metal stock for chassis fabrication.
- From the chassis layout drawings, the student will employ proper sheet metal layout techniques to dimension the actual blank stock for finished perimeter dimensions, hole locations, and bend lines.
- The student will produce the finished chassis subassemblies to include the following sheet metal fabrication skills: drilling, punching, notching, nibbling, and bending.

EXERCISE 1: Interpreting the Sheet Metal Layout Drawing

Tools & Equipment

- 1 set of chassis layout drawings

Procedure

1. Refer to figure 4-1. Note that the completed chassis consists of three subassemblies: front panel, deck plate, and back panel. Each section will be fabricated separately. The front panel will remain as an independent unit. The back panel, however, will be permanently fastened to the deck plate; together they will function as a unit.
2. Refer to the individual subassembly layout diagram shown in figures 4-2, 4-3, and 4-4 and complete Worksheet 4-1: Blank Stock Requisition Form.

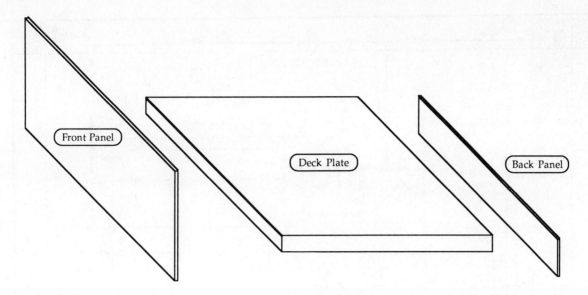

FIGURE 4-1 Chassis subassembly

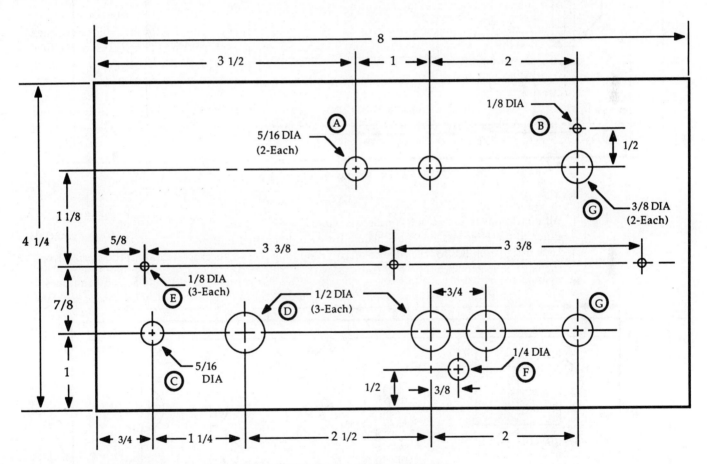

FIGURE 4-2 Front panel layout drawing

NOTE:

1. All dimensions in inches unless otherwise noted.
2. Refer to the Hole Assignment Schedule for each designated hole.

Chassis Fabrication   131

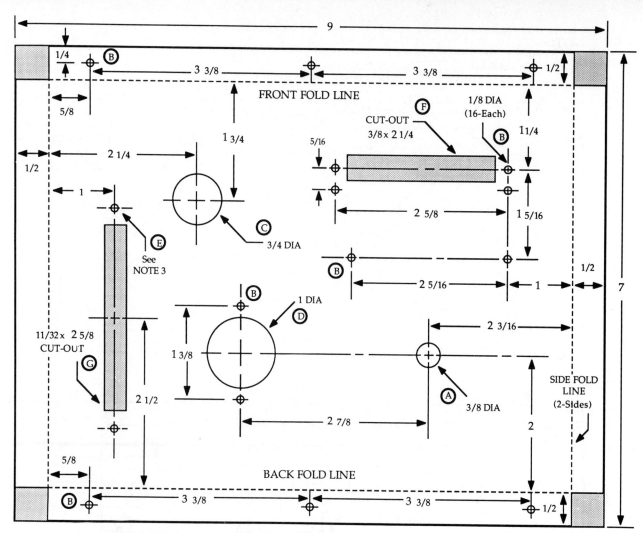

FIGURE 4-3 Deck plate layout drawing

NOTE:

1. All dimensions in inches unless otherwise noted.
2. Refer to the Hole Assignment Schedule for each designated hole.
3. Hole designation **E** is for the edge connector mounting screws. Its exact location is transferred from the edge connector positioned in cutout slot **G**.

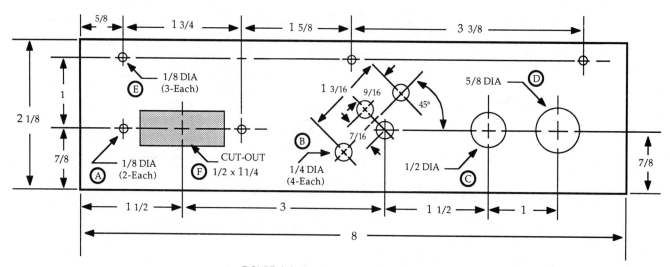

FIGURE 4-4 Back panel layout drawing

132   Prototype Construction Projects

NOTE:

1. All dimensions in inches unless otherwise noted.
2. Refer to the Hole Assignment Schedule for each designated hole.

WORKSHEET 4-1: Blank Stock Material Requisition Form

Subassembly	Length	Width	Thickness	Material
Front panel			3/32″	Aluminum
Deck plate			1/16″	Aluminum
Back panel			3/32″	Aluminum

NOTE: The recommended thickness and material is already included in this worksheet. Refer to the Appendix for the common gauge size of aluminum stock used for electronic chassis.

3. Refer to each layout diagram (front panel, deck plate, and back panel) to complete the following Hole Assignment Schedule. For each hole designated by a circled letter, indicate the correct hole dimension in the appropriate column.

NOTE: Do not fill in the column under "Produced by" at this time. This column will be completed in a later exercise.

WORKSHEET 4-2: Hole Assignment Schedule

Front Panel

Hole Design.	Assignment	Dimension	Produced by:
A	Pushbutton switches (S2 & S3)		
B	4-40 × ½″ machine screws (Fastens front panel to deck plate)		
C	Neon lamp lens cover		
D	Main power switch (S1) and output jacks (J1 & J2)		
E	Rotary switch (S4) and pot (R2)		
F	LED 1 (+5-volts dc indicator lamp)		
G	R2 10 kohm pot		

Deck Plate

Hole Design.	Assignment	Dimension	Produced by:
A	Rubber grommet		
B	4-40 × ½″ machine screw to fasten following parts: transformer (T1), terminal block (TB1), octal socket (J4)		
C	Rubber grommet		
D	Octal socket (J4)		
E	4-40 × ¾″ machine screw for edge connector (J3)		
F	Cutout for terminal block (TB1)		
G	Cutout for edge connector (J3)		

Back Panel			
Hole Design	Assignment	Dimension	Produced by:
A	4-40 machine screws to mount terminal board (TB2)		
B	TO-3 socket for IC regulator (U1).		
C	Fuse holder for line fuse (F1)		
D	Line cord with strain relief		
E	⅛″ holes for pop rivet to fasten back panel to deck plate		
F	Cutout slot for terminal board (TB2)		

EXERCISE 2: Sheet Metal Layout Technique

When the sheet metal assemblies are to be fabricated from the layout drawings, all dimensions and any other positioning information must first be transferred to the blank metal stock. Basic layout tools include the scribe, center punch, and combination square. Other specialized tools may be used to make the layout procedure more efficient. For example, the *hermaphrodite calipers* might be useful when scribing a series of parallel lines that are equidistant from the metal edge, such as those for the bend lines forming the mounting tabs on the deck plate. Regardless of what tools are used, extreme care must be practiced during layout to prevent errors.

Tools & Equipment

- Squaring shear
- Layout dye
- Sheet metal scribe
- Combination square
- Center punch
- Lightweight hammer or mallet

Optional items: Hermaphrodite calipers, steel ruler, and hole template

Procedure

1. Refer to Worksheet 4-1: Blank Stock Material Requisition Form. Obtain the proper blank stock for each subassembly as indicated in this worksheet. The length by width dimensions of each should be approximately 1″ to 2″ *greater* than its finished dimensions.
2. Using the squaring shear and with one side of the metal stock firmly against the stationary edge, shear one edge. This first cut will provide two "true" reference sides, figure 4-5. Place an identifying mark on both reference edges with layout dye, figure 4-6. Apply layout dye over the entire surface of the metal. Allow the dye to dry completely before proceeding to the next step.

FIGURE 4-5 Squaring shear to provide true sides

FIGURE 4-6 Marking true edges with layout dye

3. Using a combination square or steel ruler, measure and mark clearly the outside perimeter dimension of each subassembly, figure 4-7. Use the combination square against the reference edges and scribe lines along each marked point. Check each piece against the layout drawing to insure that the correct perimeter dimensions are present before proceeding to the next step.
4. Using the squaring shear, cut along each marked line. Each piece is now ready to be laid out according to its respective drawing.
5. Proceed to lay out the front panel first, figure 4-8. Measure and scribe the locations of all centerlines noted by C_L. Measure and scribe the center location of each hole. Scribe the designated letter beside each hole for easy identification during the drilling or punching operation. This practice of labeling each hole with its assigned letter should be used for the remaining two sections. It is not necessary to scribe the exact hole dimension for each hole. This procedure is optional but can be easily accomplished by employing a hole template. Check each hole location against the drawing before proceeding to the next step.

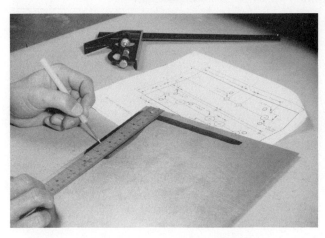

FIGURE 4-7 Combination square to measure and mark outside perimeter

FIGURE 4-8 Laying out the front panel

6. Using the center punch and a lightweight hammer or mallet, carefully and accurately punch the centers of each hole, figure 4-9. This step is very important and necessary before holes are drilled or punched
7. The deck plate will be laid out next. First, measure and scribe the ½" tabs lines along each side. Scribe an "X" at the four corners to indicate sections that will be notched out, figure 4-10.
8. Continue to lay out all circular hole locations followed by the two rectangular cutouts. Double-check all hole locations against the layout drawing before going to the final layout procedure.
9. Using the center punch, carefully and accurately punch the centers of *only* the round holes at this time.

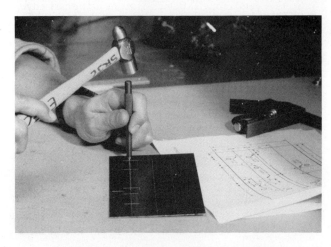

FIGURE 4-9 Center punching hole locations

FIGURE 4-10 Deck plate layout indicating corner notches

Chassis Fabrication

FIGURE 4-11 Back panel layout

10. The back panel is the most challenging section of the three to lay out and therefore has been reserved for last. First, measure and scribe the centerline (C_L) as indicated by the layout drawing, figure 4-11.
11. Measure and scribe the centers of holes C, D, and E. Measure and lay out the rectangular cutout F and associated holes A.
12. Study the layout of the four holes designated by the letter B before proceeding to lay out this hole pattern. Pay special attention to the reference hole from which all other holes are measured.
13. Double-check the location of each hole according to the layout diagram. Center punch all round holes only.

EXERCISE 3: Operating Sheet Metal Equipment to Complete Chassis Fabrication

Sheet metal fabrication employs metal cutting tools for shearing stock cleanly and precisely with a minimum amount of effort. The basic equipment for this work is the benchtop or floor model squaring shear and the 90° hand-held or benchtop notcher. Drilling with a hand-held drill motor or larger drill press is used primarily to produce small holes (less than ½″). Punching is the preferred method because the hole produced is far superior in quality than that obtained by drilling. Lever-actuated hand-held punches (Whitney), larger turret punches (Rotex), or chassis punches can all be employed to produce round holes through various gauges and material of sheet metal stock. Large square, rectangular, or irregular-shaped cutouts do not lend themselves to common drilling or punching techniques. A combination of punching followed by using a special hand nibbler will produce these irregular-shaped holes. Finally, the chassis is shaped by using a special hand-operated bending brake to prescribed angles. This particular chassis style requires only a minimum amount of bends at 90 degrees. As a result, detailed information on equipment adjustments, bend allowances, and bending sequence will not be provided but must be considered when applicable.

Tools and Equipment

- Squaring shear
- 90° notcher
- Hand-held Whitney punch set
- Rotex turret punch
- Drill press
- Drill set
- Hand nibbler
- Sheet metal brake
- Pop rivet gun with ⅛″ × ¼″ rivets
- Hand drill motor
- Metal file

Procedure

1. Refer to the Hole Assignment Schedule completed in Exercise 1. Complete the last column of worksheet 4-2 entitled "Produced by" by determining the most efficient method available to produce the variety of holes on each chassis section.
2. For the most efficient *means* to produce all holes (see figures 4-12 and 4-13), check the Hole Assignment Schedule or layout drawings directly to determine how many times the same size hole appears on each section. For example, the ⅛" holes occur in three places on the front panel; eight places on the deck plate; and five places on the back panel. Therefore all ⅛" holes should be produced on all sections before changing to the next size drill or punch.
3. After all round holes are produced on each section, the four corners on the deck plate can be notched by using the 90° corner notcher, figure 4-14.

FIGURE 4-12 Producing holes with the drill press

FIGURE 4-13 Producing holes with the hole punch

FIGURE 4-14 Notching corners with the 90° notcher

4. The deck plate and back panel contain rectangular holes. The production of these holes is completed in two steps. First, a round hole is either punched or drilled inside the cutout area. The size of this hole should be large enough to accommodate the cutting head of the hand nibbling tool, figures 4-15A and B. If one has never used this tool to produce any cutouts, practice on a scrap piece before attempting on the actual workpiece. Finish or remove any sharp edges with a metal file.

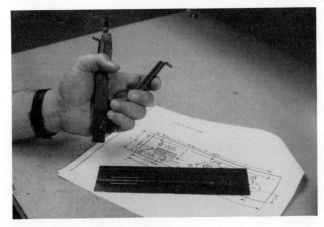

FIGURE 4-15A The hand nibbler

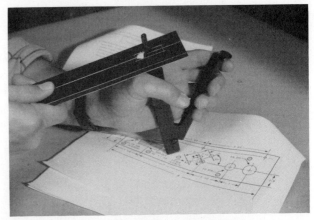

FIGURE 4-15B Producing the rectangular cutouts with the hand nibbler

Chassis Fabrication

FIGURE 4-16 Producing bends with the sheet metal brake

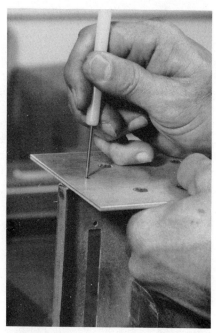

FIGURE 4-17 Layout of deck plate mounting holes with front or back panel

FIGURE 4-18 Drilling mounting holes with hand drill

5. The final bending or folding operation involves only the deck plate section. The benchtop or floor model sheet metal brake is employed in this operation, figure 4-16. Refer to the layout drawing and notice the dash lines. These are the bend lines at which 90° bends will be produced. The direction of each bend is critical, but since all bends are in the same direction, the bending sequence is not. Check the squareness of each bend with the combination square.
6. Inspect each section to assure that the finished sections conform to all requirements on the layout drawings.
7. This final step requires drilling the mounting holes on the front and back tabs on the deck plate before rivets are used to fasten the back panel to the deck plate, figure 4-17. The back panel mounting holes are produced by placing the back plate against the back tab of the deck plate and scribing each hole. Then each hole is center punched and drilled with the hand drill, figure 4-18. The back panel is then riveted to the deck plate with $1/8'' \times 1/4''$ pop rivets. The procedure for locating the mounting holes of the front panel to the deck plate front tabs is similar to the method for locating the holes on the back panel. The front panel, however, is not permanently attached to the deck plate like the back panel.
9. The basic chassis for Project 1 is now completed with the exception of the chassis finishing procedure. This procedure is optional and will be left to the decision of the individual student or instructor.

NOTE: *The following assignments, Project 1 and Project 2, introduce basic assembly techniques and were not designed to be a student take-home project. The chassis is recycled after the completion of Project 3, the wiring procedure. In fact, most of the parts assembled to this chassis in Projects 2 and 3 are salvaged and reused by the next group of students. As a result, an attractive finished (painted) chassis will not retain its new look after repeated use. I, therefore, would not suggest applying a finish to this chassis unless this is a student take-home project of which he or she can be proud of.*

NOTE: *Figure 4-19A-D illustrates how the chassis will look once the indicated parts have been mounted.*

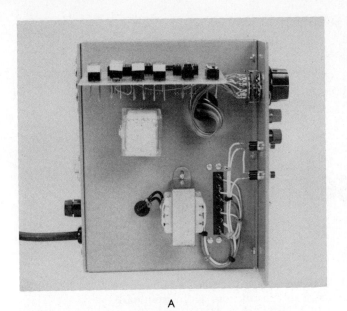
A

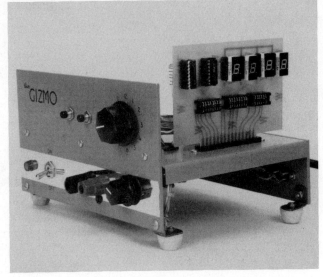
B

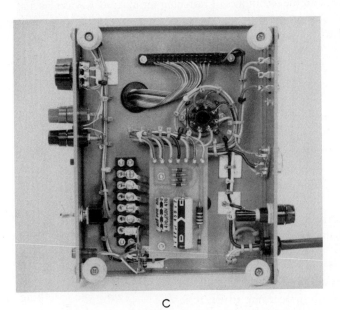
C

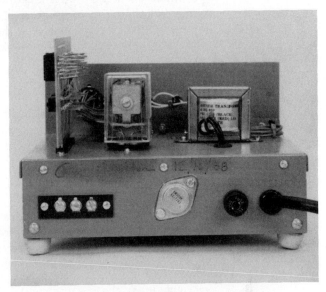
D

FIGURE 4-19 Completed chassis: (A) top view, (B) front panel view and side view showing test card, (C) wiring view, (D) rear view: showing low voltage ac output, IC reg. fuse and line cord. Also seen are the components on the deck plate: test card, relay, and transformer.

Chassis Fabrication   139

Project 2: Chassis Assembly

PURPOSE

- To develop the ability to interpret written assembly instructions and illustrations.
- To introduce common electronic hardware and the techniques of fastening terminating devices such as terminal blocks, strips, and sockets.
- To prepare the chassis assembly for the wiring procedure.

COMPONENTS & EQUIPMENT

- Two sections of prefabricated aluminimum chassis
- Miscellaneous hardware and components (see Table 4-1)
- Basic hand tools, a soldering iron, and miscellaneous tools

EXERCISE 1: IDENTIFICATION OF HARDWARE AND COMPONENTS

The purpose of this exercise is to identify the hardware and components that will be mounted on the chassis. Identifying each item by its proper name will make the assembly procedure easier.

The parts list is provided a checklist to insure that all component parts are present before proceeding to the assembly procedure and to provide a means of identifying any component substitutions the technician may encounter as a result of design changes.

Carefully locate and identify each part before proceeding to Exercise 2.

TABLE 4-1 Parts Identification and Checklist

ITEM NO. (CHECK ✔)	DESCRIPTION	ILLUSTRATION OF PART
1	Two prefabricated chassis	
2	Rectifier board (assembled in Assignment 6) *(continues)*	

140 Prototype Construction Projects

ITEM NO. (CHECK ✔)	DESCRIPTION	ILLUSTRATION OF PART
3	Rotary switch (SP-10 position) with knob and mounting hardware (round washer and hex nut)	
4	10-kilohm pot with knob and mounting hardware (lock washer and hex nut)	
5	Two push-button switches: red (PBNO) and black (PBNC)	
6	Red and black banana jacks with mounting hardware (two no. 10 hex nuts and two no. 10 solder lugs)	
7	LED indicator lamp, red lens with mounting clip, and ring (*continues*)	

Chassis Assembly 141

ITEM NO. (CHECK ✔)	DESCRIPTION	ILLUSTRATION OF PART
8	Toggle switch (SPST)	
9	Neon lamp with lens and retainer clip	
10	15-pin edge connector	
11	8-pin octal socket	
12	6-terminal barrier strip	
13	4-lug terminal strip	
14	Transformer: filament type; 12.6 volts ac; 1 amp

(continues) | |

ITEM NO. (CHECK ✔)	DESCRIPTION	ILLUSTRATION OF PART
15	Line cord: 3 conductor; 16 gauge; 3 prong	
16	Line cord strain relief	
17	Fuse holder: chassis type with mounting hardware (rubber washer, lock washer, and nut)	
18	LM309K, 5-volt regulator in TO-3 package; TO-3 style transistor socket with mounting hardware (mica washer, two no. 6, ½" sheet-metal screws	
19	Screw-type, 3-terminal, chassis-mount terminal strip with mounting hardware (two 6-32 × ¼" machine screws with lock washers and hex nuts)	
20	Rubber grommets: one ¼" IC; one ½" ID (*continues*)	

Chassis Assembly 143

ITEM NO. (CHECK ✔)	DESCRIPTION	ILLUSTRATION OF PART
21	Insulated stand-offs: two no. 4 × 1" long; one no. 6 × ¼" long	
22	Machine screws and nuts For: Item 2, the rectifier board, and item 14, the transformer	Size: Two 4-40 × ½" screws, four hex nuts, and one no. 4 solder lug for each
	Item 10, the 15-pin edge connector	Two 6-32 × ½" screws, two hex nuts, and two lock washers
	Item 11, the 8-pin octal socket	Two 6-32 × ¼" screws, two hex nuts, and two lock washers
	Item 12, the 6-terminal barrier strip	Four 4-40 × ¾" screws, four hex nuts, and four lock washers
	Item 13, the 4-lug terminal strip	One 6-32 × ½" screw only

EXERCISE 2: CHASSIS ASSEMBLY PROCEDURE

The chassis assembly procedure is divided into three sections: the front panel, the base panel, and the back panel. The first step in the assembly procedure is to fasten the two chassis sections together (see figure 4-20).

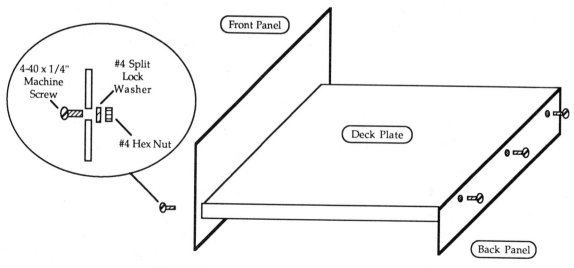

FIGURE 4-20 Initial chassis assembly procedure

144 Prototype Construction Projects

NOTE: *Two 4-40 × ¼" machine screw, a no. 4 split lock washer, and a no. 4 hex nut are required to completely fasten the chassis sections together.*

The three sections of the chassis are shown in figure 4-21. This figure will be used throughout the assembly procedure to mount the components in the proper locations. Note that the number assigned to each round hole or rectangular slot in this figure is the same as the number in the Item Number column in the Parts Checklist completed in Exercise 1.

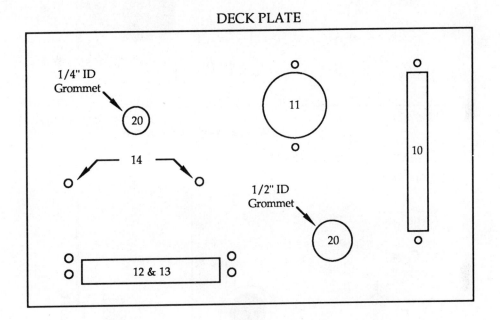

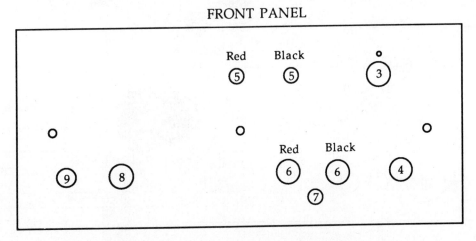

FIGURE 4-21 View of the three major chassis sections

Chassis Assembly 145

Mount the components on the chassis. Figures 4-22 through 4-32 are provided to aid in the assembly process.

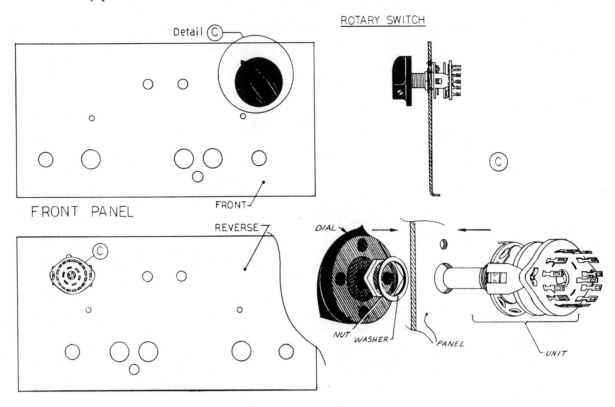

FIGURE 4-22 Item 3, rotary switch

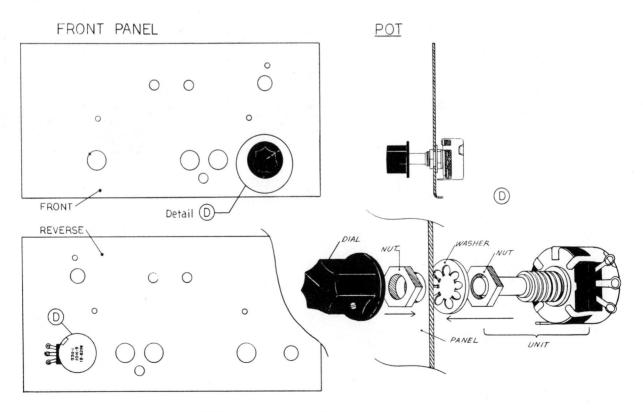

FIGURE 4-23 Item 4, pot (flashing rate control)

146 Prototype Construction Projects

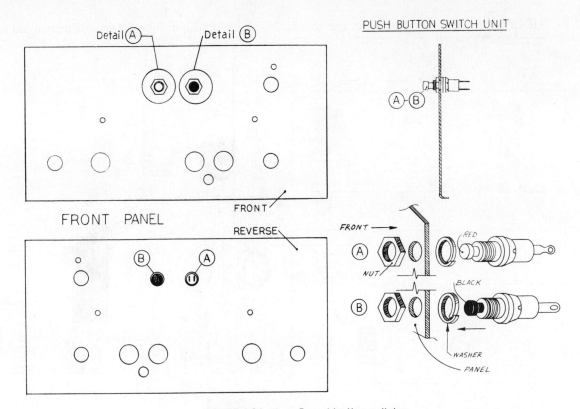

FIGURE 4-24 Item 5, pushbutton switches

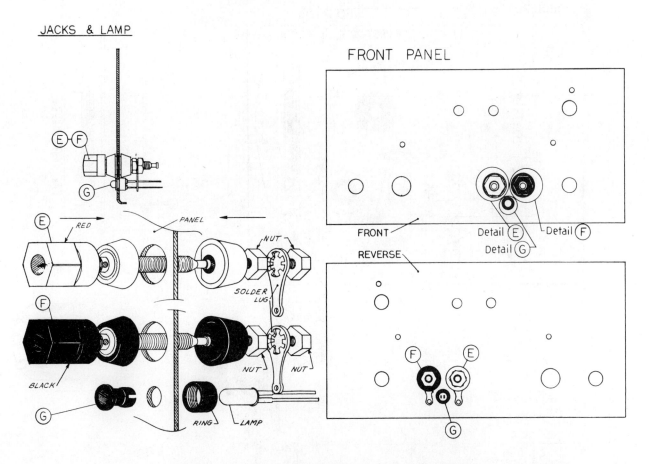

FIGURE 4-25 Item 6, banana jacks; item 7, LED indicator lamp

Chassis Assembly 147

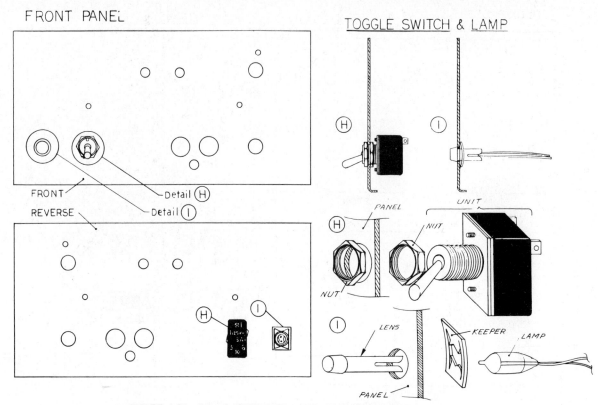

FIGURE 4-26 Item 8, SPST toggle switch; item 9, neon lamp with lens

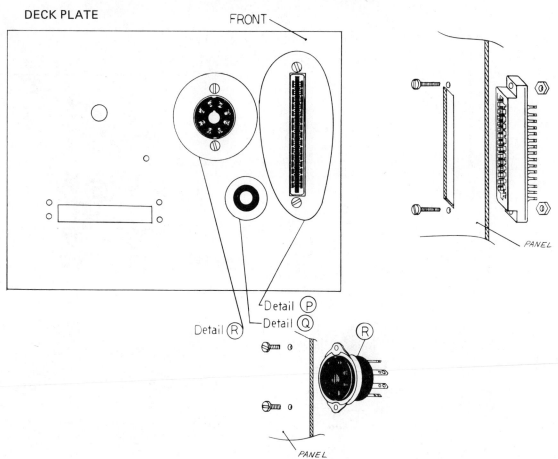

FIGURE 4-27 Item 10, 15-pin edge connector; item 11, 8-pin octal socket; item 20, rubber grommet, ½" ID

148 Prototype Construction Projects

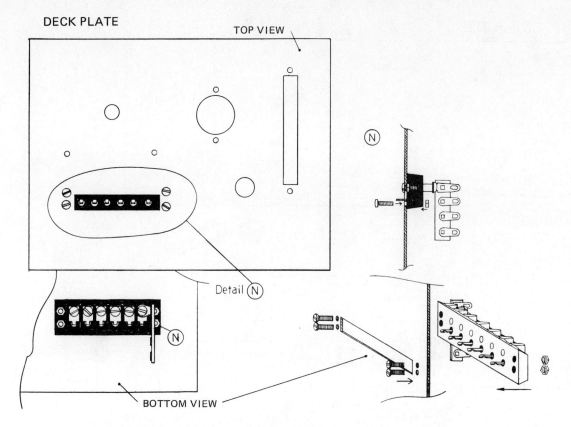

FIGURE 4-28   Item 12, 6-terminal barrier strip; item 13, 4-lug terminal strip

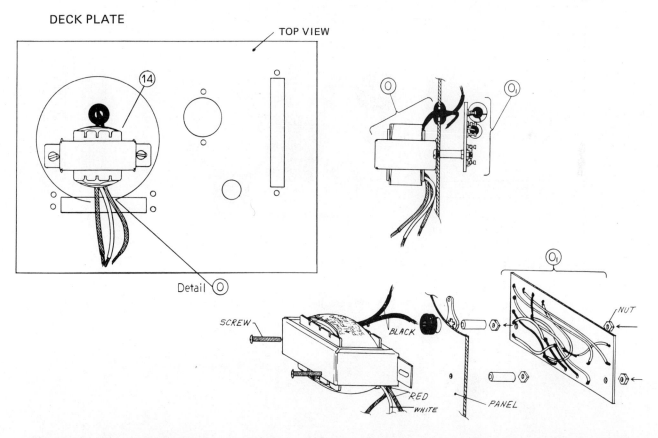

FIGURE 4-29   Item 14, filament transformer; item 20, ¼" rubber grommet; item 21, two number 4 × 1" insulated standoffs; item 2, assembled rectifier board

Chassis Assembly   149

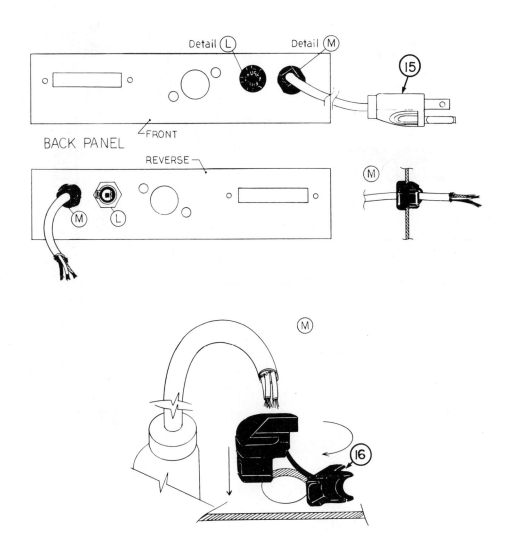

FIGURE 4-30 Item 15, line cord, item 16, line cord strain relief

150　Prototype Construction Projects

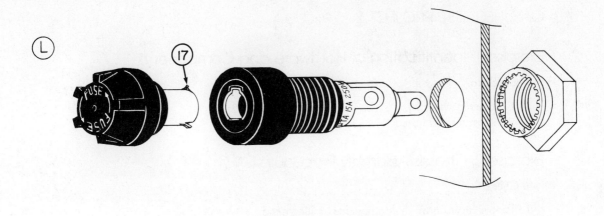
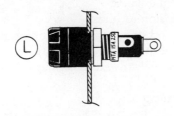

FIGURE 4-31 Item 17, fuse holder

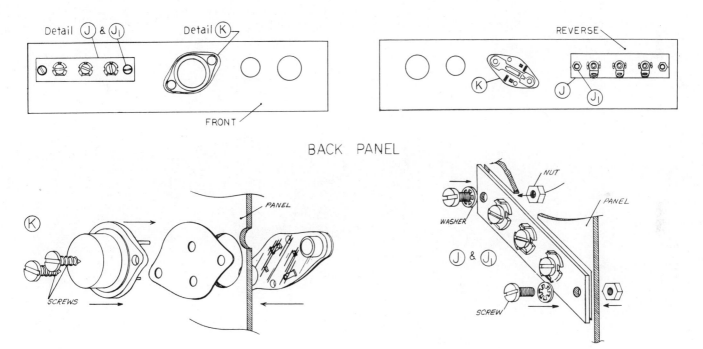

FIGURE 4-32 Item 18, LM309D 5-volt regulator; item 19, screw-type terminal strip

Chassis Assembly 151

QA CHECKLIST: PROJECT 2

Exercise 1: Identification of Hardware and Components

Check for:

1. Completed checklist
2. All substitutions noted

Exercise 2: Chassis Assembly Procedure

Check for:

1. Proper location of all components as illustrated
2. Proper orientation of rectifier board
3. Proper orientation of transformer leads
4. Proper orientation of barrier strip, edge connector, octal socket, and terminal strips
5. Proper assembly of LM309k 5-volt IC regulator
6. Correct sequence of fastening hardware as illustrated (combination of washer, lock washer, and nut combination)
7. Proper torque on all fastening hardware

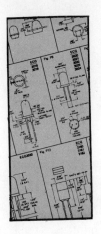

Project 3: Chassis Wiring Procedure

PURPOSE

- To introduce four common wiring diagrams: the block diagram, the schematic diagram, the connector wiring diagram, and the chassis wiring diagram.
- To wire the chassis prepared in Project 2, Exercise 2.
- To develop a connector wiring assignment table or chart from a given wiring diagram.
- To utilize basic test equipment as a multimeter to confirm the electrical specifications of the unit and to perform basic troubleshooting procedures such as voltage and continuity measurements.
- To develop schematic reading skill by performing the wiring procedures directly from the schematic diagram.
- To display proper assembly techniques such as solid and stranded wire preparation, crimping, and soldering, which were developed in Section 3.

COMPONENTS & EQUIPMENT

- Assembled chassis unit
- Wire assortment
- Relay (RL1) and edge connector test board

NOTE: These two items will be used only when required during certain phases of the wiring procedure.

EXERCISE 1: IDENTIFICATION OF WIRING DIAGRAMS

Wiring diagrams are used by people who design, assemble, and maintain equipment. Each type of diagram has a specific function. For example, an engineer who is changing existing electrical circuitry would refer to the schematic diagram before initiating any changes. If changes were being made to systems other than the electrical system, the engineer might use the block diagram, the connector wiring diagram, or the chassis wiring diagram.

The service technician who maintains the equipment would use the entire set of diagrams to perform periodic maintenance on the equipment and to insure that all engineering changes are performed as instructed.

The assembler would use the connector wiring diagram, the chassis wiring diagram, and any detailed diagrams that might simplify complex assembly operations. The assembler probably would not use the block diagram or the schematic diagram.

The Chassis Wiring Diagram

Inspect the chassis wiring diagram which shows the top and bottom views of the assembled and wired unit. Notice the orderly paths of wire bundles (harness) between the terminating points.

NOTE: *Do not attempt to wire your chassis with this diagram only. The wiring procedure will be explained after the orientation to the diagrams.*

DECK PLATE (TOP VIEW)

DETAIL C

DETAIL A DETAIL B

154 Prototype Construction Projects

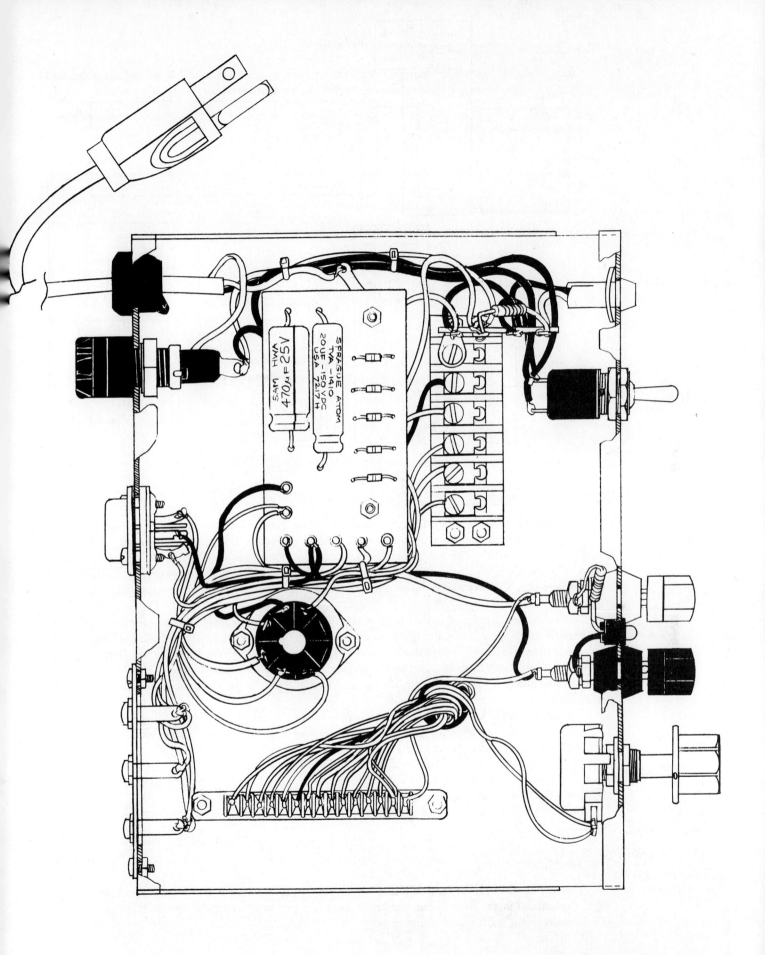

Chassis Wiring Procedure 155

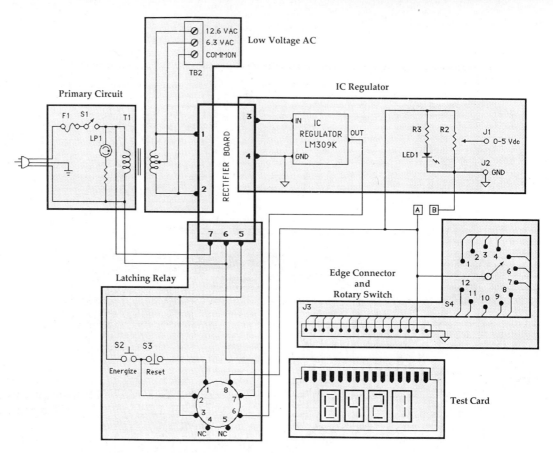

FIGURE 4-33 Block diagram of test fixture

The Block Diagram

The orientation for the block diagram and the schematic diagram is a programmed guide that requires one to respond immediately to a question or to perform a specific task before progressing to the next activity. As each block is discussed, questions relating to components or other details in that block are asked. The questions and activities are designed to walk one through the diagrams. It is, therefore, imperative that you answer each question correctly or perform each activity before progressing to the next step.

Notice the names assigned to the blocks in the block diagram of the test fixture (see figure 4-33). The first portion on the left side of the diagram is called the primary circuit. It consists of the line voltage which is represented by its symbol, followed by the fuse (F1), the main switch (SW1), and the lamp circuit which consists of an indicator lamp in series with a resistor. What type of indicator lamp is this?

The line isolation follows the primary circuit. What component provides the important function of line isolation?

The block entitled low voltage ac source is a convenient source of low ac voltage that can be obtained from this test fixture. Unfortunately, there are only two *fixed* values of low ac voltage. Notice that the line isolation *transformer* is directly supplying both ac output voltages.

Locate the transformer (T1) on your chassis and identify the primary and secondary leads. The two black leads are the primary leads and the remaining three wires are the secondary leads. What are the colors of your transformer's secondary leads?

Indicate the color and location of your transformer's primary and secondary leads directly on the block diagram.

The Rectifier Board

The rectifier board is the heart of this test fixture in that it functions to insure the proper operation of all of the sections in the fixture. The numbered points on this block represent the turret terminals. The number assignment is illustrated in figure 4-34.

156 Prototype Construction Projects

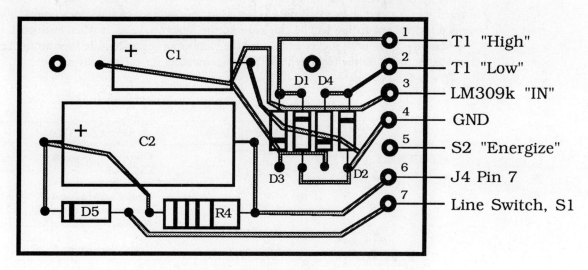

FIGURE 4-34 Rectifier board terminal number assignment

The rectifier board consists of two separate rectifier sections. The exact components and the rectifier circuit configurations are not very apparent in this block representation, but each type will be discussed in detail in the schematic diagram section.

The outputs of this rectifier board are provided by terminals 3, 4, and 5. What is the symbol at the output of terminal 4 (figure 4-33)?

There is another symbol identical to this one at output jack J2. What is the significance of these two identical symbols?

The voltage at terminal 3 along with the output at terminal 4 feeds the input of the following block, the IC regulator block. Another output voltage is present at terminal 5. The available output voltage at this point is approximately 120 volts dc.

CAUTION: *Do not measure this voltage at any time with any test equipment that must be plugged into the same line voltage as this test fixture. Use a VOM for this voltage measurement.*

The output voltage from terminal 5 feeds push-button switches S2 and S3. What type of push-button switch is S2 (*energize function*)? What type of push-button switch is S3 (*reset function*)?

IC Regulator

As was mentioned in the discussion on the rectifier board, the output voltage at terminal 3 feeds the IC regulator section. This IC (integrated circuit) regulator is a three terminal device. Locate this device on your chassis.

Carefully examine this device and notice how it is mounted via a special socket on the back panel. The case of this device is the *GND (ground)* terminal. Locate this terminal on the socket.

Locate the input and output terminals on the same socket.

What is the case style of this packaged unit?

Latching Relay Circuit

The latching relay circuit includes both push-button switches, S2 and S3, and relay RL1. The electrical rating of this relay is:

- Coil Voltage—approximately 100–150 volts dc
- Coil Resistance—200 ohms
- Contact—DPDT, 5-amp maximum

Chassis Wiring Procedure   157

A relay is simply an electromagnetic switch. When the proper coil voltage is provided, the coil is energized which causes it to move the contact arms to *close* a set of stationary points. The actual relay provided for this test fixture will be plugged into the 8-pin octal socket when instructed.

Locate the 8-pin octal socket and notice the numbering sequence in relation to the keyed slot. The number assignments on the block represent the pin numbers for the relay as follows:

PIN NO.	RELAY SECTION
1	Contact arm, section A
2	Coil
3	Normally-open contact A
4	Normally-closed contact A
5	Normally-closed contact B
6	Normally-open contact B
7	Coil
8	Contact arm, section B

The electrical action of the relay is not apparent unless the internal sections of the relay are known. The schematic diagram provides this information; therefore, it is not required on the block diagram.

The output of the IC regulator section feeds pin 6 of the relay as illustrated on the diagram. Pin 8 is then terminated to J1. What is J1?

J1 also provides a fixed regulated dc voltage of 5 volts. What does the 1-amp rating represent?

Locate J1 on the chassis. What color is this output jack? What color is J2?

What is the purpose of the series combination of LED 1 and resistor R3?

15-Pin Edge Connector

The edge connector test board is inserted into the 15-pin edge connector to provide a convenient means of testing the sequence of events for each position of rotary switch S4. Locate the edge connector and observe how the contacts are designed to connect with the fingers of the PC board when inserted into this connector.

Notice the numbering sequence for the pins (contacts). Looking at the connector from the side into which the PC-board card is to be inserted, where is pin 1 in relation to the front panel?

Rotary Switch

The rotary switch shown in this diagram is a *shorting* rotary switch. What does this indicate?

Locate this switch on the front panel and rotate the shaft while observing the contact action of the arm to the various contacts. Is it evident now why it is called a shorting rotary switch?

Another type of rotary switch is the *non-shorting* rotary switch. What does this mean?

This rotary switch is also identified as a single-pole rotary switch. What is meant by this description?

Rotary switches are also available in a variety of poles and positions. For example, 2-pole, 6-position; 3-pole, 3-position, etc. What is meant by these examples of poles and positions?

Viewing this rotary switch from the rear, a *clockwise* rotation of the shaft will cause the positions (terminals) to be numbered in a *counter-clockwise* direction. Which view of the numbered positions is represented on the block diagram for rotary switch S4—front or rear view?

EXERCISE 2: SCHEMATIC READING AND THE WIRING PROCEDURE

The wiring procedure is divided into five sections: the primary circuit, the latching relay, the low voltage ac source, the IC regulator, and the rotary switch and edge connector. Each major section may be wired with the aid of any available diagram; however, it is encouraged that the schematic diagram be consulted to determine the electrical connections between components within each section. Written wiring procedures are required for the primary circuit, the latching relay, and the IC regulator sections. The entire wiring procedure is provided for the primary circuit only. A partial procedure is provided for the latching relay section and none is provided for the IC regulator section. The major objective of the written procedures is to provide a means of reading the schematic diagram or other diagrams.

The Primary Circuit

The primary circuit is illustrated on the left side of the schematic diagram of this test fixture (see figure 4-35). Identify and locate all components that make up this circuit.

The primary circuit consists of the line voltage, the switch (S1), the fuse (F1), and the series combination of the neon lamp (LP1, NE2) and the resistor (R1 = 68 kilohms). The primary winding of transformer T1 is also part of the primary circuit. Although diode D5 is electrically connected to a number of components in the primary circuit, it is not included in this section.

The primary circuit will be the first section wired. Before proceeding to the next wiring exercise, this section must be inspected by your instructor. The wiring instructions that follow are presented in tabular form. The reason for this method of instruction is to introduce a method of documenting a wiring procedure.

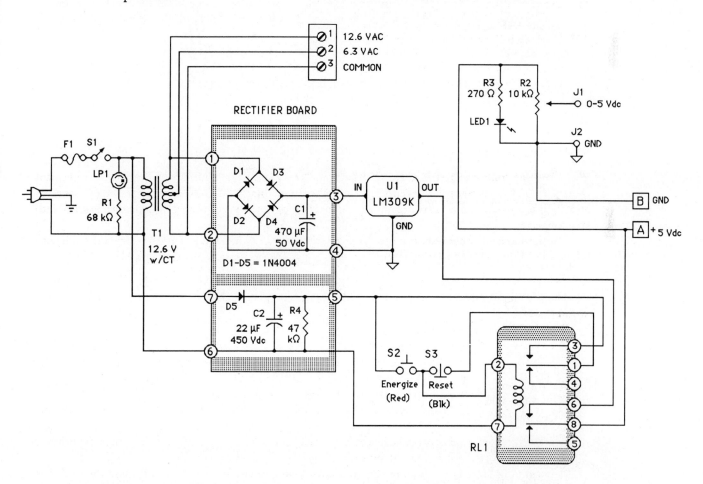

FIGURE 4-35 Schematic diagram of test fixture

TABLE 4-2 Primary Circuit Wiring Procedure

NOTE: *All wire terminations that require crimping may need more than one wire connected to a single terminal. Do not solder any crimped terminals until instructed.*

STEP	INSTRUCTION	NOTES
1.	Line cord. Black wire (hot) to terminal 1 of fuse holder.	Solder.
2.	Connect no. 22 stranded black wire from fuse holder terminal 2 to any terminal of S1.	Solder.
3.	Connect no. 22-gauge black wire from the unsoldered terminal of S1 to TS1-3.	TS1-3 = terminal strip 1, terminal 3. Solder terminal S1 only.
4.	Connect one black lead of transformer T1 (primary) to TS1-3.	
5.	Connect no. 22-gauge black wire from TS1-3 to turret terminal (TT) 7 on the rectifier PCB.	Solder wires on TS1-3 and TT7.
6.	Connect the neutral, white wire of line cord to TS1-2.	TS1-2 = terminal strip 1, terminal 2.
7.	Line cord green lead to the solder lug mounted directly on the metal chassis. This lug may be inserted on any mounting screw that makes direct contact with this chassis.	Solder.
8.	Insulate both leads of neon lamp LP1 with spaghetti or insulation removed from approximately 22-gauge solid or stranded wire.	
9.	Connect one lead of the neon lamp to TS1-4 and the second lead to TS1-3.	TS1-4 = terminal strip 1, terminal 4. TS1-3 = terminal strip 1, terminal 3.
10.	Connect R1, the 68 kilohm, ½ W resistor, between TS1-4 and TS1-2.	
11.	Connect 22-gauge white wire from TS1-2 to turret terminal 6 of the rectifier PCB.	
12.	Connect the second black lead of T1 to TS1-2.	Solder.

The wiring procedure for the primary circuit is now completed. *Do not plug the circuit into line voltage until instructed.*

Double-check your wiring for:

1. Incorrect wire locations
2. Improper crimps
3. Cold-soldered joints
4. Too much solder
5. Wire route too taut

Using *only* the schematic diagram, trace the actual point-to-point wiring given in the wiring procedure.

Have the wiring for this section inspected and approved by your instructor before proceeding to the next wiring exercise.

Plug the line cord into the line voltage receptacle. Turn S1 on. The neon lamp should light.

The Latching Relay

From this point on, only the schematic diagram of the test fixture (Figure 4-35) is used to complete the wiring procedure.

The latching relay circuit incorporates semiconductor diode D1, electrolytic capacitor C2, pushbutton switches S2 and S3, and relay RL1. Using Table 4-2, write the component values for this section directly on the schematic diagram.

Locate components D5 and C2. These components should be mounted on the rectifier board.

Locate pushbutton switches S2 and S3 and relay RL1.

Switches S2 and S3 should have been mounted on the chassis front panel. Identify the normally-open and the normally-closed pushbutton switches by the color of the buttons.

Relay RL1 will be plugged into the octal socket, which is located on the chassis base section.

Locate the relay on the schematic diagram and note the assigned numbers around the symbol. What do these numbers represent? What do the dotted lines around the symbol represent?

NOTE: Before inserting the relay into the octal socket, have the wiring technique inspected and approved by your instructor.

The input to this section was completed in the writing procedure for the primary circuit. Can you determine from the schematic diagram which two connections were completed in that wiring procedure?

The anode of component D5 is connected to TT-7 (turret terminal 7) and the negative side of C2 is connected to TT-6. Make a notation of your schematic diagram for these two points.

The note beside turret terminal number 5 (TT-5) in figure 4-34 indicates that TT-5 is "to S2 & pin 3, J4." Locate this point and make a notation directly on your diagram showing the location.

The input to this section begins from rectifier PCB terminals 6 and 7. The relay will be plugged into the octal socket that is responsible for providing all electrical connections to the actual relay.

Pushbutton switches S2 and S3 are located on the chassis front panel. The function of S2 is to activate, or energize (energ.), the relay coil to complete the switching action of the DPDT switch section within the relay housing. One half of the DPDT switch is terminated on pin numbers 1, 3, and 4. The second half is connected to pins 5, 6, and 8. Pins 1 and 8 of the relay are the wiper, or common arm, of the relay contacts. Pins 3 and 6 are the NO (normally open) contacts and pins 4 and 5 are the NC (normally closed) contacts. Notice that pins 4 and 5 are not connected (NC).

The function of S3, the PBNC (pushbutton, normally-closed) switch, is to de-energize or reset the relay. TT-5 on the rectifier board is also connected to this section as noted—to S2 & pin 3, J4.

The wires from the pushbutton switches located on the top-side must be routed and terminated to the turret terminals on the rectifier board and octal socket pins located on the bottom side.

Locate TB1, the terminal block that consists of six screw terminals on the bottom side and lug terminals on the top side. Note that barrier strip TB1-1 is also electrically connected to TS1-1. To provide a convenient terminating point to complete the wiring of both pushbutton switches to the bottom-side connection points, only terminals 1 and 2 of TB1 will be used.

The same method of documenting a wiring procedure will be used as in the primary circuit wiring procedure. The wiring instructions, however, will be provided by each student. Following are suggested steps for the wiring procedure.

WORKSHEET 4-2: Latching Relay Section Wiring Procedure

NOTE: Relay pins 6 and 8 will not be terminated in this section. They will be connected in the voltage regulator section.

STEP	INSTRUCTION	NOTES
1.	Rectifier board TT-5: Connect no. 22-gauge wire from TT-5 to pin 3, J4 (octal socket).	
2.	Connect the second wire on TT-5 to TB1-2 (screw terminal side).	

Complete the remaining instructions.

STEP	INSTRUCTION		NOTES
3.	S2:	Collect no. 22 gauge wire on terminal-lug side of TB1-2. Other end of wire to one terminal of S2 (red, PBNO).	Solder both terminals.
4.	S3:		
5.	Relay pin 7:		
6.	Relay pin 1:		

Before applying voltage to this section, have your wiring procedure inspected and approved by your instructor.

The initial testing procedure involves listening to the relay "click" when S2 (red button) is depressed, and to the "clock" when S3 (black button) is depressed. The red pushbutton switch should energize the relay circuit and the black pushbutton switch should de-energize the circuit.

The next test displays the operational characteristics of the relay circuit. An ohmmeter is required for this test. Perform and record the resistance measurements for *only* those pins indicated in Worksheet 4-3.

CAUTION: Resistance measurements taken with the ohmmeter across relay pins other than those indicated in Worksheet 4-3 may damage this instrument.

WORKSHEET 4-3: RLY1 Contact Continuity Test

PIN NUMBERS BETWEEN	RELAY CONDITION	RESISTANCE INDICATION
8 and 6	De-energized	
8 and 5	De-energized	
5 and 6	De-energized	
8 and 6	Energized	
8 and 5	Energized	
5 and 6	Energized	

The Low Voltage Ac Source

The wiring procedure for this section involves terminating the secondary leads of T1 to terminal block 2 (TB2) located on the chassis back panel. Before performing this wiring procedure, note that the terminals of terminal block TB2 are located on the inside of the chassis while the secondary leads of T1 are located on the top side of the chassis. The secondary leads of T1 must be terminated to the barrier strip, TB1. Connect each secondary lead of T1 and TB1 as follows:

1. First red lead to TB1-4
2. White lead (center tap) to TB1-5
3. Second red lead to TB1-6

Using the instructions that follow, complete the wiring procedure for this section. Before performing this final wiring procedure, answer the following question:

Why was it necessary to terminate the secondary leads of T1 to terminal block TB1 lug terminals numbers 4, 5, and 6?

ANSWER: *The convenience factor. Since the three secondary leads were located on opposite sides of their final destination point (terminal block TB2), the terminals of TB1 being available on both sides of the chassis provided a convenient means of getting T1's secondary leads to the inside of the chassis. TB1 also provides a rigid support for these leads. Another convenience is the availability of the secondary leads to service personnel to perform voltage or resistance measurement of the T1 secondary winding.*

The final wiring instruction for the low voltage ac source follows:

1. Red wire from TB1-4 (screw terminal) to TB2-1
2. White wire from TB1-5 (screw terminal) to TB2-2
3. Orange wire from TB1-6 (screw terminal) to TB2-3

After completing this wiring procedure, this section will be tested for proper operation. The test for this section is to measure the ac voltage across the output terminals of TB2 with an ac voltmeter. Complete Worksheet 4-4 for each output voltage measurement as indicated in the table.

WORKSHEET 4-4: Measured Output Voltage of the Low Voltage Ac Source Section

TEST POINTS WITH RESPECT TO TERMINAL 3	MEASURED VOLTAGE (RMS VALUE)
1	
2	

The IC Regulator Section

Using Worksheet 4-5, each student must provide the wiring instructions for the IC regulator. The instructions may be documented as in the previous sections. Use only the schematic diagram to complete the wiring.

The IC regulator section consists of:

1. +dc output voltage from the rectifier board—junction of D3 and D4
2. Filter capacitor C1
3. IC regulator U1
4. Relay RL1 contacts on pins 6 and 8
5. Series combination of LED 1 and R3
6. Voltage-adjust pot, R4

The +5 volts dc output voltage provided from this section is connected to one outside terminal of R2 (potentiometer). The center terminal is the wiper for the pot. A wire must be soldered to this terminal and to the red jack. The black jack, J2, which is also located on the front panel, provides the ground (GND), or common, return for this section. A wire must be soldered to this jack and to the third terminal of the pot.

Before attempting to wire this section, locate all components and terminating points. For example: A reading of the schematic diagram indicates that the +dc voltage from the bridge rectifier at the junction of D3 and D4 and the positive (+) side of capacitor C1 feeds the input (IN) terminal of the IC regulator, U1. By reading further, we see that diodes D3 and D4 and capacitor C1 are mounted on the rectifier PCB.

The logical sequence of reading, or interpreting the electrical blueprint is not an easy task. The relationship of the actual physical component mounted on the chassis or on the PCB to the schematic diagram is not obvious in many instances. Keep one important concept in mind as you are attempting to read the schematic to produce the written wiring procedure: every component must have *at least two leads*, or sides. When performing the connection from a terminal or component to another point on the diagram, complete *all* connections to that side of the component before attempting to complete the other side.

WORKSHEET 4-5: Wiring Procedure for the IC Regulator Section

STEPS	INSTRUCTION	NOTES
1		
2		
3		
4		
5		
6		
7		
8		

To test the IC regulator section for proper operation:

1. Obtain RL1 and insert it in the octal socket.
2. Turn main switch S1 on.
3. Depress S2, the red pushbutton switch, to energize the relay. If the IC regulator is wired properly, the LED should turn on.
4. With the voltage-adjust pot rotated fully clockwise, measure the dc voltage at J1 with respect to ground (J2). The voltage should measure approximately 5 volts dc. A counterclockwise rotation should decrease the voltage from 5 volts to 0 volts.
5. Depress S3 to reset RL1. The LED should turn off and there should be no voltage at the output jack, J1.

To troubleshoot the IC regulator:

1. Check the LED for proper polarity or defects.
2. Check the wiring to the relay contacts: pins 6 and 8.
3. Be sure that U1 is wired correctly—input, output, and ground terminals on socket.
4. Check the ground return wires for missing wires.
5. Be sure that there is no +dc output voltage at terminal 3 of the rectifier board.

The Rotary Switch and Edge Connector

The rotary switch and edge connector section is the final wiring section for this test fixture. The wiring of this section will be performed using only the Edge Connector and Switching Diagram shown in figure 4-37. Before beginning the wiring process, turn to figure 4-35, the schematic diagram of test fixture. Note the letters "A" and "B" within the square symbols. These symbols also appear on the Edge Connector and Switching Diagram. Refer to Figure 4-36.

Locate rotary switch S4 and edge connector J3 on the chassis. What type of rotary switch is S4? (Describe its poles and positions.)

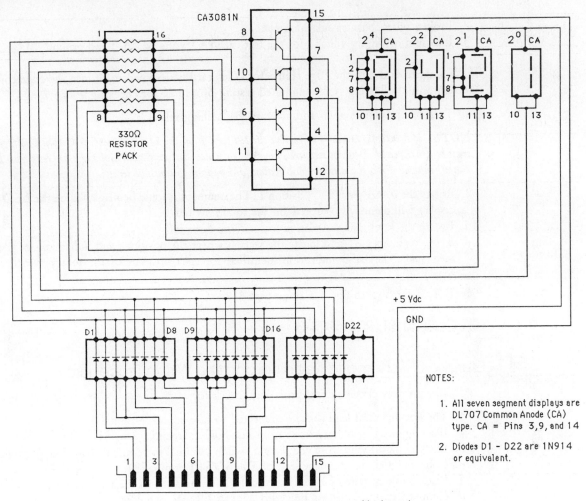

FIGURE 4-36 Schematic diagram of test card

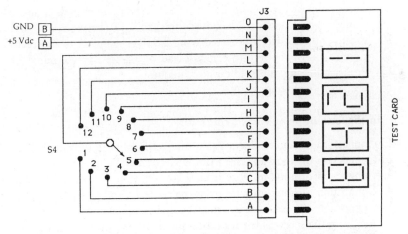

FIGURE 4-37 Schematic diagram of rotary switch and edge connector

Edge connector J3 is a 15-pin edge connector which is located on the base of the chassis. It accepts a special test card the purpose of which is to provide a means to test for correct wiring.

A ribbon cable will be used to complete the termination between S4 and J3. Note the wire colors that make up this ribbon cable. Does the color sequence follow a logical or standard color/number order (brown (#1), red (#2), orange (#3), yellow (#4))? It may be desirable to consider this color/number sequence when wiring the switch or edge connector.

Using only the diagram shown in figure 4-37, wire this section.

Before using the test card, check for:

1. Solder bridges between the terminals of the edge connector or the switch terminals

Chassis Wiring Procedure   165

2. Broken wires, burnt insulations, etc.
3. Inclusion of supply voltage for the test card (+5 volts dc "A") and ground "B"

Insert the card into the edge connector. This card is designed to fit into the edge connector in one direction only. Following is the suggested testing procedure for checking your wiring:

1. Turn on the main switch and activate the latching relay.

NOTE: If it becomes necessary to remove the test card from the edge connector, the applied voltage to this card must be disconnected. It is not required, however, to switch the main switch off. Use the reset push button (black button) instead. Push the activate push button (red button) to continue service.

2. Rotate the rotary switch to position 1. The number *1* should be displayed on the LED readout. This display will identify position 1 on the rotary switch.
3. Rotate the knob one position in a clockwise direction.
4. Refer to figure 4-37 and locate the section where the positions and indications of the test card are identified. The LED indication identified for each position of the rotary switch must occur in sequence.

NOTE: Refer back to figure 4-19 on page 139.

QA CHECKLIST: PROJECT 3

Exercise 2: Schematic Reading and the Wiring Procedure

In the primary circuit wiring procedure, check for:

1. The hot line cord lead (black) connected to fuse-holder, terminal 1
2. Black hook-up wire connected from fuse-holder, terminal 2 to S1
3. Black hook-up wire from the second switch terminal to terminal strip TS1 (terminal 3)
4. Connection of one primary lead of T1 to TS1 (terminal 3)
5. Black hook-up wire (22 gauge) from TS1 (terminal 3) to the rectifier board (terminal 7)
6. Connection of the neutral line cord wire to TS1 (terminal 2)
7. Connection of the ground lead of the line cord (green) to the terminal lug
8. Connection of neon lamp LP1 to TS1 (between terminals 4 and 3)
9. Connection of resistor R1 (68 kilohms) to TS1 (between terminals 2 and 4)
10. White hook-up wire connected from TS1 (terminal 2) to the rectifier board (terminal 6)
11. Connection of the second primary lead of T1 to TS1 (terminal 2)
12. Proper crimping and soldering techniques
13. Proper wire routing

In the latching relay circuit, check for:

1. Connection of hook-up wire from the rectifier board (terminal 5) to the octal socket, J4 (pin 3)
2. Connection of hook-up wire from TS1 (terminal 1) to the rectifier board (terminal 5)
3. Connection of hook-up wire from TB1 (terminal 1) to the first terminal of S2 (PBNC)
4. Connection of hook-up wire from the second terminal of S2 to the first terminal of S3 (PBNO)
5. Connection of hook-up wire from the first terminal of S3 to the solder lug side of TB1 (terminal 2) and from the screw side of TB1 to pin 2 of J4 (octal socket)
6. Connection of hook-up wire from the second terminal of S2 to the solder lug side of TB1 (terminal 3) and from the screw side of TB1 to pin 1 of J4
7. White hook-up wire connected from the rectifier board (terminal 6) to pin 7 of J4
8. Proper crimping, soldering, and wire routing techniques
9. Proper resistance readings in Worksheet 4-2, the RLY1 contact continuity test:

- Pins 8 & 6 de-energized = Infinite
- Pins 8 & 5 de-energized = Zero
- Pins 8 & 6 energized = Zero
- Pins 8 & 5 energized = Infinite

In the low voltage ac source, check for:

1. Connection of one secondary lead of T1 to the solder lug side of TB1 (terminal 4) and from the screw side of TB1 to TB2 (terminal 1)
2. Connection of the center-tapped lead of T1 to the solder lug side of TB1 (terminal 5) and from the screw side of TB1 to TB2 (terminal 2)
3. Connection of the second secondary lead of T1 to the solder lug side of TB1 (terminal 6) and from the screw side of TB1 to TB2 (terminal 3)
4. Proper crimping, soldering, and wire routing techniques
5. Proper voltage measurements at TB2 (Worksheet 4-3):
   - Terminal 1 = Approximately 14 volts ac
   - Terminal 2 = Approximately 7 volts ac

In the IC regulator section, check for:

1. Connection of hook-up wire from the rectifier board (terminal 3) to the *in* terminal of U1
2. Connection of one hook-up wire from the rectifier board (terminal 4) to the ground terminal of U1 and of the second hook-up wire from either terminal (terminal 4 on the rectifier board or the ground terminal of U1) to J2 (black binding post)
3. Connection of hook-up wire from the *out* terminal of U1 to J4 (terminal 6, octal socket)
4. Connection of the black hook-up wire from the *X* terminal of J4 to J2
5. Connection of the red hook-up wire from J4 (terminal 8) to S2
6. Connection of the series combination of LED 1 and R3 across J1 and J2, with the anode of LED 1 connected to J1 and R3 connected to J2
7. Proper crimping, soldering, and wire routing techniques
8. Proper operation of this section: When S2 (red PBNO) is depressed, relay RL1 is energeized, and LED 1 is turned on. A clockwise rotation of the voltage-adjust control increases output voltage from 0 to 5 volts dc and +5 volts dc appears at J1 (measured with respect to ground, J2). When S3 (black PBNC) is depressed, RL1 is de-energized, LED 1 is turned off, and there are zero volts at J1.

In the rotary switch and edge connector section, check for:

1. Connection of black hook-up wire from any ground point to J3 (contact 1).
2. Connection of red hook-up wire from J1 to the high side of R2 and to the arm of S4 (rotary switch).
3. Connection of red hook-up wire from the high side of R2 or the arm of S4 to J3 (contact 2).
4. Connection of hook-up wire between the low side of R2 and the wiper and from this point to J3 (contact 3).
5. Connection of ribbon cable from J3 to S4 as follows:

J3 CONTACTS	TO	S4 TERMINAL
A		1
B		2
C		3
D		4
E		5
F		6
G		7
H		8
I		9
J		10
K		11
L		12
M		wiper
N		+5 Vdc
O		GND

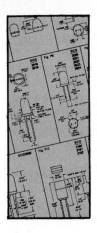

Project 4: Continuity/Voltage Tester

The continuity/voltage tester is used to test two electrical quantities—resistance and voltage. It is equivalent to an analog or digital multimeter (VOM) which is capable of measuring specific values of voltage, resistance, and current. This tester, however, is limited by the fact that it is only an indicator of voltage or resistance; it does not measure current. The major limitation of this tester, therefore, is its inability to display specific values of voltage or resistance, a task which an analog or digital multimeter does perform.

The continuity function of this tester is designed to detect any "opens," or "shorts," in a number of electrical or electronic applications. A single conductor in a multiconductor cable can easily be isolated with the aid of this tester. Any open (infinite resistance) conductors in a multiconductor cable can be found, as can conductors that develop shorts (zero resistance) within the cable housing. Specific elements of electronic devices, such as diodes or transistors, can also be located.

Although the continuity/voltage tester does not display specific values of voltages, it has two ranges—low and high. The low range indicates on the LED a voltage range of 3 volts to 50 volts. The high range indicates on the neon lamp a voltage range from approximately 55 volts to 200 volts. This tester also determines whether the unknown voltage is ac or dc.

The fabrication procedure for this project is illustrated in the block diagram shown in figure 4-38.

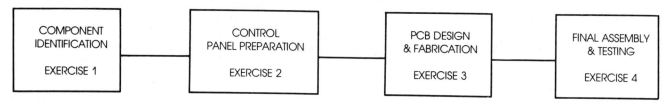

FIGURE 4-38 Block diagram of Project 2 construction schedule

Each block represents an exercise that must be performed, evaluated, and approved before progressing to the next exercise.

EXERCISE 1: COMPONENT IDENTIFICATION

Purpose

- To correctly identify the components represented on the schematic diagram.
- To identify the components that will be mounted on the PCB.
- To identify the components that require external wires or leads and that will not be mounted on the PCB.

Components & Equipment

- Component kit for tester

Procedure

Using the schematic diagram of the continuity/voltage tester shown in figure 4-39, identify each component in your parts kit from its symbol on the schematic diagram. After each component is correctly identified, place the actual component directly on the appropriate symbol on the diagram.

NOTE: *Refer to the Quick Reference Guide in Appendix A to aid in the identification process.*

Note that some components provided in the parts kit are not represented on the schematic diagram by a symbol. These items are considered miscellaneous and may be accessory items such as battery snaps, a neon lens cap, a battery holder, etc.

Using the checklist which follows, complete the inventory of all components and miscellaneous items provided in the kit.

PARTS CHECKLIST

Components

___ R1   1 kilohm, 5 %, 1-watt carbon resistor
___ R2   220 ohms, 5 %, ½-watt carbon resistor
___ R3   82 kilohms, 5 %, ¼-watt carbon resistor
___ D1   1N4004, 1-amp, 400 volt diode
___ LED1   Jumbo, red light-emitting diode
___ LP1   Ne2, neon lamp
___ SW1   DPDT slide switch
___ SW2   SPDT slide switch
___ B1   9-volt transistor battery

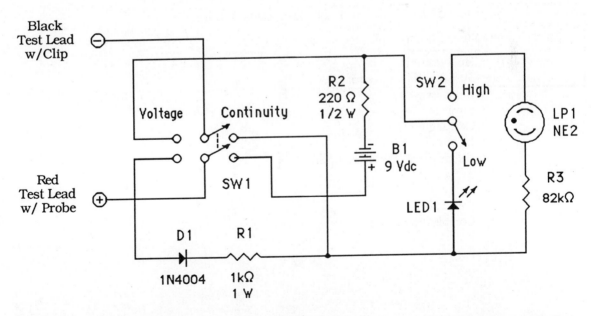

FIGURE 4-39 Schematic diagram of continuity and voltage tester. The function switch, SW1, is in the continuity position; the range switch, SW2, is in the low-range position.

Miscellaneous Items

___ Red test probe

___ Battery holder

___ Battery snap

___ LED mounting clip

___ Neon lamp lense cap

___ Plastic case

___ ⅜" I.D. rubber grommet

___ Alligator clip with black boot

___ 4 machine screws and 4 hex nuts 4-40 × ¼"

EXERCISE 2: PREPARATION OF THE CONTROL PANEL

Purpose

- To identify individual components or miscellaneous items that will be mounted on the control panel.
- To interpret a mechanical drawing (working drawing) of the control panel.
- To perform layout techniques on the control panel using the working drawing.
- To introduce basic construction skills such as punching, drilling, and other hole producing techniques to prepare the control panel for assembly.
- To complete the assembly of the control panel and to breadboard (make an experimental arrangement of) the entire continuity/voltage tester circuit.

Tools & Equipment

Materials

- 5" × 5" single-sided or double-sided blank PCB material

Drafting Equipment

- Hole template
- 30/60 or 45-degree triangle
- Engineer's scale
- T-square
- Drafting board
- Light table
- Graph paper (10 squares/inch)

Sheet Metal Layout Tools

- Steel scale
- Combination square

- Scribe
- Center punch
- Hammer (small claw or ball-peen)
- Drill gauge
- Hand files
- Tapered reamer

Equipment

- Drill press (bench or floor)
- Hand-held hole punch
- Turret punch
- Squaring shear (bench or floor)
- Miscellaneous twist drills

Procedure

Using the schematic diagram in figure 4-39, circle the following symbols on the diagram:

1. Black clip
2. Red probe
3. SW1
4. SW2
5. LED1
6. LP1

The purpose of circling these symbols is to identify only those components or accessory items that will be mounted directly on the control panel plate. The control panel plate will be fabricated from the blank PCB material.

Remove the circled items from the parts kit and set them aside. Instructions to determine correct hole dimensions and other important dimensions for each of these items will be provided in the following steps.

In addition to the circled items, locate the ½" I.D. rubber grommet for use in this exercise.

The blank PCB serves two functions. It provides a mounting surface for all switches, indicating lamps, and other miscellaneous items for the tester. At the same time, its "inside" surface accommodates the remaining components on a printed circuit which will be designed and fabricated in the following exercise.

The control panel layout diagram (figure 4-40) indicates the relative location of each hole with respect to the top, the left edge, and the center line of the panel. The holes are identified in Table 4-3.

TABLE 4-3 Hole Location and Identification

HOLE	IDENTIFICATION
A	Neon lamp lens cap
B	LED mounting cap
C	Function switch (SW1)
D	Range switch (SW2)
E	Panel mounting holes
F	SW1 and SW2 mounting holes

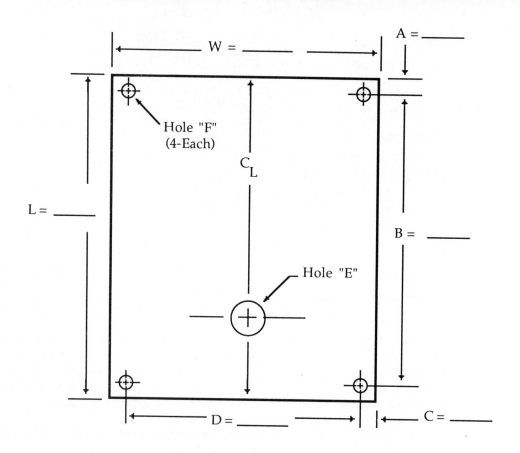

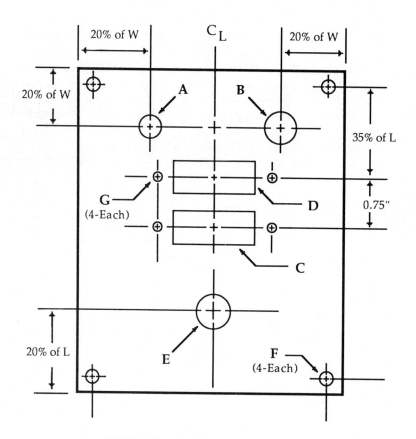

FIGURE 4-40 Control panel layout diagram

FIGURE 4-41 Determination of height and width of control panel

The dimensions represented by the letters *W* and *H* must be determined first. A method of obtaining these dimensions is shown in figure 4-22. Write your measurements directly on the diagram in figure 4-39 (*W* = _____ and *L* = _____). These two dimensions will serve as a basis of determining the exact location of all of the holes on the panel.

The next step is to develop the actual working drawing. Use one sheet of graph paper (10 squares/inch) for this exercise. Your working drawing should indicate the exact location and size of each hole. The hole locations are determined by calculating the percentages of dimensions *W* and *H* indicated in figure 4-39. For example: if W = 4″ and H = 5.5″, the location of hole *A* is approximately 1.1″ from the top edge (20% of 5.5″ = 1.1″). The location of holes *C*, *D*, and *E* with respect to the top edge of the panel can be determined in the same manner.

The symbol C_L stands for center line. This dimension, which is 50% of *W*, can easily be determined by taking one half (½) of *W*.

Use a hole template to locate the panel's four mounting holes (holes *F*). These can be determined with the scale.

Once each hole location is known, the diameters of the holes must be determined. It is important to represent the *exact* dimensions of each hole directly on the working drawing. Methods of determining hole dimensions for each item on the control panel are shown in figures 4-42 through 4-44.

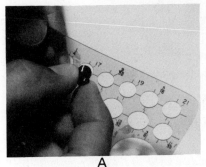

A   B

FIGURE 4-42 (A) LED1 mounting clip, (B) LP1 lend cap

FIGURE 4-43 Rubber grommet

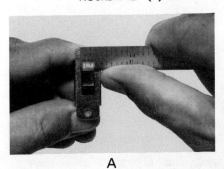

A

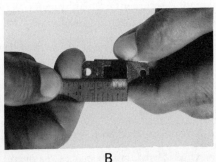

B

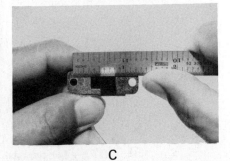

C

FIGURE 4-44 SW1 and SW2: (A) width of slot, (B) length of slop, (C) mounting holes

Continuity/Voltage Tester 173

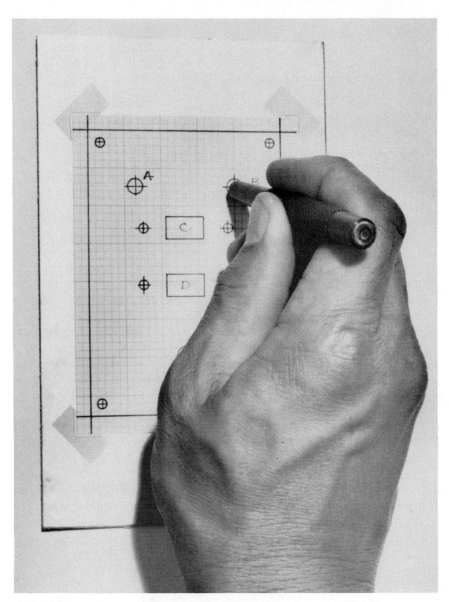

FIGURE 4-45 Working drawing taped on PCB blank and hole locations transferred with center punch

After completing your working drawing, trim it and tape it to the blank PCB material. This layout diagram will serve as your hole template to transfer hole locations directly to the PCB blank (see figure 4-45).

NOTE: *If single-sided PCB material is used, the working diagram should be taped on the copper-foil side.*

When your layout diagram is securely fastened, transfer all hole locations directly to the PCB blank by center-punching all hole locations as illustrated in figure 4-45.

NOTE: *The four corners of the control panel should also be center-punched. These points, however,* **will not** *be drilled. Their purpose is to provide reference points to scribe perimeter cutting lines after the layout template is removed.*

NOTE: *Do not remove the working drawing from the PCB blank at this time.*

Complete Worksheet 4-6 and proceed to Exercise 3, PCB Design and Fabrication.

NOTE: *The operations required to produce each hole on the control panel will not be performed at this time. This will be covered in the following exercise.*

WORKSHEET 4-6: Hole Production Schedule

Complete the table below by indicating the specific hole-producing equipment and the proper size of the punch or drill in the appropriate column.

HOLE	IDENTIFICATION	EQUIPMENT	PUNCH OR DRILL SIZE
A	Neon lamp lens cap (LP1)		
B	LED mounting cap (LED1)		
C	Function switch (SW1)		
D	Range switch (SW2)		
E	Rubber grommet		
F	Panel mounting holes		
G	SW1 and SW2 mounting screw holes		

EXERCISE 3: PCB DESIGN AND FABRICATION

Purpose

- To identify all components that will be soldered on to the PCB.
- To design the printed-circuit trace network from a component position diagram.
- To fabricate the PCB using the direct-etch process.
- To assemble the completed PCB and to interface with items on the control panel.

Tools & Equipment

For PCB Design

- Graph paper (0.1" grid)
- X-acto knife
- 1:1 PCB artwork patterns: 1/8" O.D. donut pattern, 1/16" tape

For PCB Fabrication

- Single-sided or double-sided blank copper-clad board
- Copper etching solution (Ammonium persulfate or ferric-chloride)
- Sheet metal shear
- Small bastard file assortment (flat, round, and triangular style)
- Soldering iron, solder, and hand tools for assembly

Procedure

Using the schematic diagram of the continuity/voltage tester (figure 4-39), draw a rectangular box around the following component symbols: D1, R1, R2, and R3. These four components (one diode and three resistors) will be mounted directly on the PCB. The PCB, therefore, will be designed around these four components and will eventually interface with the corresponding items mounted on the control panel.

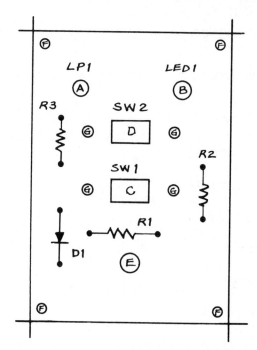

FIGURE 4-46 Component placement on PCB

Remove these four components from the parts kit and examine their *sizes* (DO-7 case, wattage, etc.), their *shapes* (rectangular, tubular, oval, etc.), and their *lead styles* (axial or radial). The sizes, shapes, and lead styles of these components are important considerations in the second phase of PCB design.

The exact physical location of these four components has been selected for this project only. The relative position of each component with respect to holes that will house LED1, LP1, SW1, and SW2 is shown in figure 4-46.

The working drawing of the hole locations, which was produced in Exercise 2, will be used in this procedure.

Since all four components are *axial* components, the following lead-forming requirements apply to all of them.

1. The lead shall extend from the body of the component a *minimum* distance equal to the diameter of the lead before the start of the bend (see figure 4-47).

    NOTE: L must be equal to or greater than D.

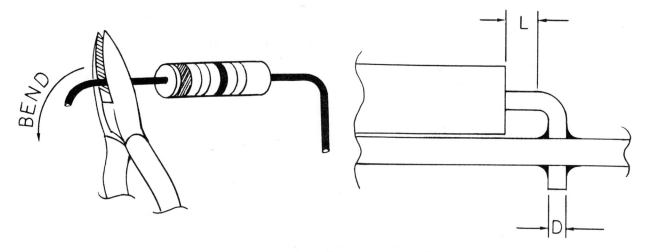

FIGURE 4-47 Lead bend requirements

176 Prototype Construction Projects

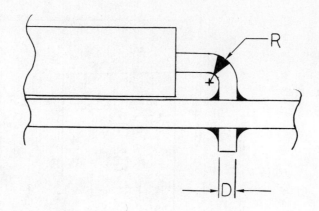

FIGURE 4-48 Bend radius requirement

2. The bend radius must be no less than ½ the lead diameter (see figure 4-48).

NOTE: R must be greater than ½ D.

Place each component at its approximate location on the working drawing and mark it with a dot to locate the exact lead location of each component. Both leads from LP1 (the neon lamp) and both leads from LED1 must also be considered at this time.

Refer to figure 4-49 for the suggested placement of the leads for LP1 and LED1. One lead of LP1 (neon) is connected directly to terminal 1 of SW2 (range), while the other is connected to the donut pad on the PCB. The cathode lead of LED1 is connected directly to terminal 3 of SW2.

CAUTION: Avoid placing components too close to the edges of the control panel. Leave approximately 0.2" from the component body to the edge of the panel. Also, avoid placing the components or leads too close to any hole. Leave approximately 0.2" from the lead position to the outside edge of any hole.

Transfer all of the lead locations on your working drawing to the PCB blank. Using the center punch, gently punch each dot as indicated on your working drawing (see figure 4-50).

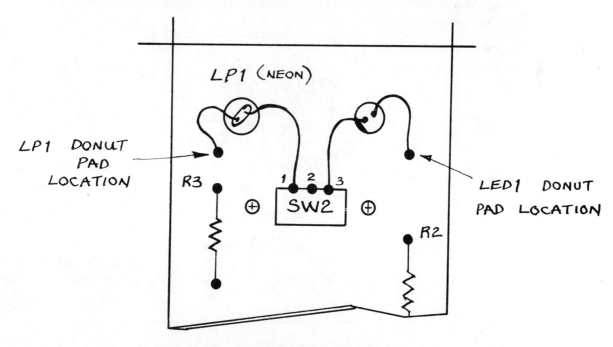

FIGURE 4-49 Lead placement for LP1 and LED1

Continuity/Voltage Tester   177

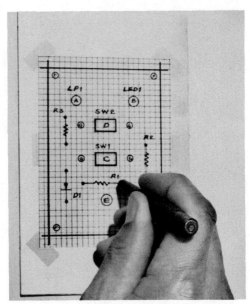

FIGURE 4-50 Transferring component lead locations to PCB blank

Remove the graph paper from the PCB blank and inspect all of the center-punched marks to insure that all points are visible. Re-punch if necessary. Donut pads will be applied directly over these points in a later step.

Note that only component leads were transferred to the PCB blank in the previous step. Look at the schematic diagram (figure 4-39) and notice that several other dots must be included to complete the lead locations. The following external leads must be considered (see figure 4-51 for suggested locations for these leads):

1. The wire lead from terminal 4 of SW1 to the *anode* of D1.
2. The wire lead from terminal 3 of SW1 to the junction of R1, LED1 (anode), and R3.
3. The black (negative) lead of the battery snap to R2.
4. The wire lead from terminal 1 of SW2 to LP1.
5. The wire lead from terminal 3 of SW2 to LED1 (cathode).

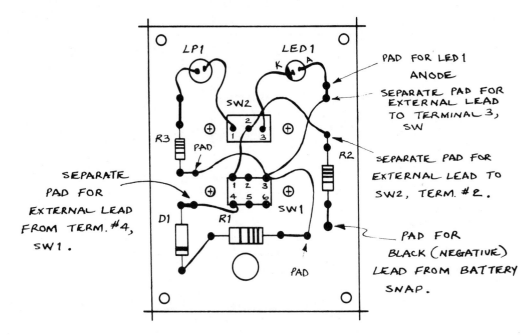

FIGURE 4-51 Suggested locations for external leads

178 Prototype Construction Projects

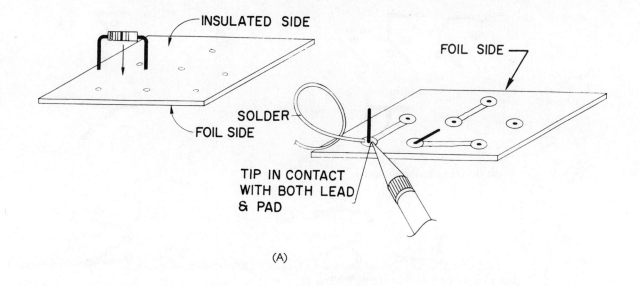

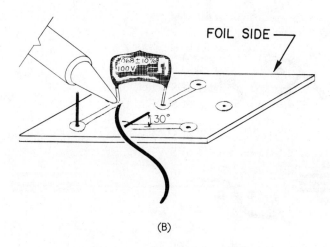

FIGURE 4-52 PCB assembly techniques: (A) traditional method, (B) special method

Providing a separate *pad* (the spot where all component leads and external wire leads are fed through the component side and soldered on the foil side) for each external lead is important when designing any standard single-sided PCB. If separate pads are not provided for external leads, the unacceptable practice of soldering directly to component leads or attempting to force two leads into one pad is common.

Providing a separate pad for this particular PCB is a requirement regardless of the soldering technique employed for this project; that is, all components will be mounted and soldered on the foil side. The traditional method of applying components to PCBs and the special technique that will be used in this project are illustrated in figure 4-52.

Use the schematic diagram (figure 4-39) to produce the *trace diagram*. Draw the conductive paths (traces) between components on your graph paper sketch and have it inspected before proceeding to the tape-up process. Examples of recommended and not recommended drafting practices for designing trace runs are shown in Table 4-4.

TABLE 4-4   Recommended Practices for Trace Runs

RECOMMENDED	NOT RECOMMENDED
1. Avoid sharp external angles. They may cause foil delamination. 	
2. Avoid acute internal angles.	
3. Use shortest practical trace routing. 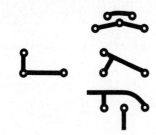	
4. Maintain equal spacing when conductors pass between pad areas. 	
5. Avoid large multiple pad areas that may cause soldering problems—that is, non-symmetrical solder fillets. 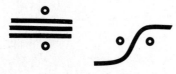	
6. Maintain uniform patterns around holes to produce symmetrical solder fillets. 	
7. Avoid using tape that is the same size as the pads. It may cause solder to flow away from the pads.	

180   Prototype Construction Projects

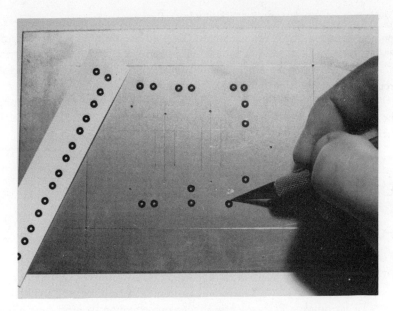

FIGURE 4-53 Application of donut pads to hole locations

Have your trace diagram sketch inspected and approved before progressing to the next step.

Using the following procedure, apply the donut pads to each hole location indicated on the PCB blank (see figure 4-53).

1. Remove pad from backing and apply to hole location.
2. After the pads have been applied to all points, place a clean sheet of paper over the pads. Rub the pads firmly with a blunt tool to insure good pad adhesion to copper foil.

Apply *PC tape* between the donut pads as illustrated in the trace diagram sketch (figure 4-46). This diagram has two functions: it provides the pad location for each component lead; and it is used to sketch the lines that represent traces between pads. The recommended technique for applying this tape is shown in figure 4-54.

NOTE: *Donut pad holes must not be covered with tape.*

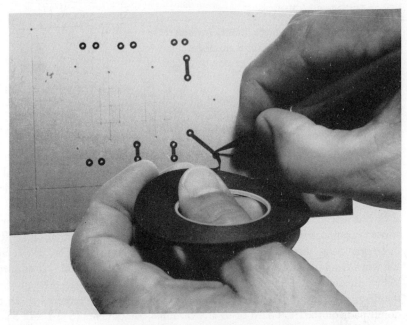

FIGURE 4-54 Application of PC tape

Continuity/Voltage Tester

Your circuit board is now ready for the final PCB fabrication process, the *etching process*. Commercial or home-made etching equipment can be used for this simple method of PCB fabrication. A properly etched board will produce a copper-foil pattern that is an exact duplication of the taped copy.

Using Worksheet 4-6, which was completed in Exercise 2, perform the required operations to reproduce each hole at its respective location.

Trim the completed control panel board with the sheet-metal squaring shear to fit the plastic chassis box. Use the flat files to finish the edges. Have the completed control panel board inspected and approved before beginning Exercise 4.

EXERCISE 4: FINAL ASSEMBLY AND TESTING

Purpose

- To introduce a method of dressing a simple control panel.
- To complete the mechanical and electrical assembly of all components on the control panel circuit board.
- To introduce a PCB soldering technique common to prototype model construction.
- To develop electrical specifications for the continuity/voltage tester.

Tools & Equipment

Tools

- Basic electronic hand tools
- Miscellaneous soldering tools

Test Equipment

- VTVM or VOM
- Decade resistance box
- Low-voltage power supply
- High-voltage power supply

Miscellaneous Materials

- Flat white spray paint (enamel or lacquer)
- Rub-on transfer letters and numbers
- Clear spray fixative

Procedure

A simple but effective method of dressing a panel will be used in this project. The panel is spray painted and then labeled with rub-on transfer letters. The entire process is illustrated in figure 4-55. After completing the process, spray a thin coat of spray fixative over the panel front to prevent the transfer letters from peeling or cracking.

The assembly procedure of all items on the control panel will be performed next (see figure 4-56 for correct component placement). All of the PCB components (R1, R2, R3, and D1) and the interconnecting wires or leads from LP1 (the neon lamp), LED1, SW1, SW2, the test probe wire, and the wires from the battery snap will be soldered. The *tack method* will be used to solder these components. When the tack method is used, component leads are not inserted through the lead locations as is normally the case. Instead, all component leads, as well as any wire required to be terminated, are tacked to the donut pads on the PC board or to any terminal components (SW1 and SW2). The technique of tacking wires or component leads is illustrated in figure 4-57.

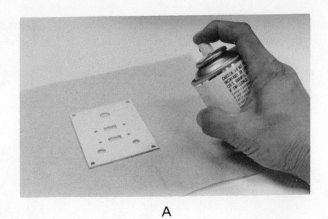

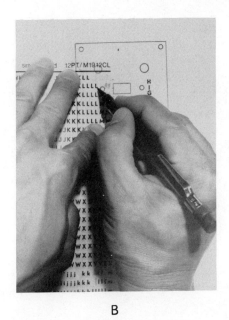

FIGURE 4-55 Process of dressing panel front: (A) spray paint, (B) transfer letters and numbers, (C) suggested titles

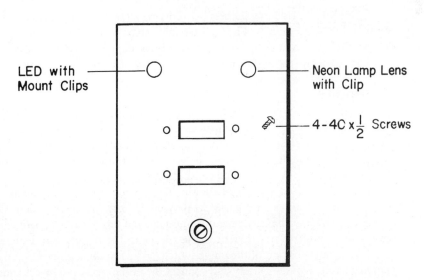

FIGURE 4-56 Control panel assembly drawing

Continuity/Voltage Tester

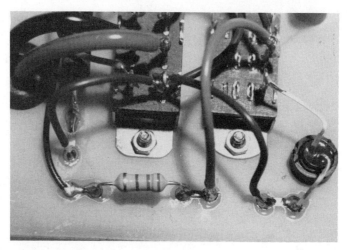

FIGURE 4-57 Tack soldering technique: (A) to PC boards, (B) to switch terminals

Complete the assembly process by soldering all of the components and interfacing the wires to the PC board and switch terminals. The final assembly step is to terminate the red test lead wire to a red test probe and the black wire to an alligator clip with a black boot (see figure 4-58).

The assembly process is now complete. Use the following procedures to determine whether the completed project is working properly and use the troubleshooting techniques in Table 4-5 to correct any assembly errors.

Connect the 9-volt battery to the battery snap. The range switch (SW2) must be on the low position and the function switch (SW1) must be on the continuity position. Connect the red test probe to the black alligator clip. LED1 should turn on. Disconnect the test probe and the alligator clip. LED1 should now turn off. The continuity function of the tester is working properly if the readings for LED1 reflected the on and off conditions for the open and short tests.

Remove the 9-volt battery from the battery snap. Move the function switch (SW1) to the voltage test position. The range switch (SW2) should still be at the low position. Connect the alligator clip to the negative terminal of the battery and connect the red probe to the positive battery terminal. LED1 should turn on for a positive test. *Reverse* the test leads to the battery terminals. LED1 should not turn on.

Change the position of the range switch (SW2) to the high position. Connect the alligator clip to the negative terminal of the battery and connect the probe to the positive terminal. *Neither* LED1 or the neon lamp (LP1) should turn on.

184 Prototype Construction Projects

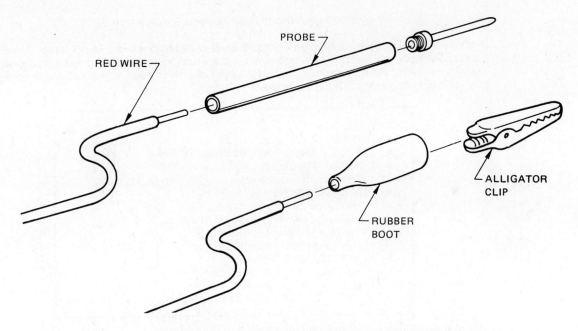

FIGURE 4-58 Test probe and alligator clip terminating technique

Leave the range switch on the high position. Locate the high-voltage, dc power supply and connect the alligator clip to the negative output terminal of the supply and the red probe to the positive output terminal. *Slowly* increase the output voltage from the supply and watch the neon lamp (LP1). The neon lamp should turn on when approximately 55 to 60 volts dc is applied to the test leads. *Do not increase the voltage any further at this point.*

If the preceding operational tests were positive for both the continuity and voltage functions, you may skip the troubleshooting techniques described in Table 4-5.

TABLE 4-5 Troubleshooting Techniques

SYMPTOM	POSSIBLE SOLUTION
CONTINUITY FUNCTION: NO LED INDICATION	1. Weak battery—replace. 2. LED not soldered properly; anode and cathode reversed—remove and reverse leads. 3. Defective LED—replace. 4. Check the value of R2 (220 OHMS)—a larger value will reduce the brightness of the LED. 5. Defective function switch (SW1)—test and replace, if necessary. 6. Defective range switch (SW2)—test and replace, if necessary. 7. Unsoldered or broken wires. 8. Cold soldered joints.
VOLTAGE FUNCTION: NO LED INDICATION ON LOW POSITION	1. D1 reversed—desolder and change position. 2. D1 defective—replace. 3. R1 defective or of incorrect value—test and replace. 4. LED1 defective—replace. 5. Defective function switch (SW1)—test and replace. 6. Defective range switch (SW2)—test and replace. 7. Unsoldered or broken wires. 8. Cold soldered joints.
NO INDICATION ON HIGH POSITION	1. Defective neon lamp (LP1)—replace. 2. Value of R3 (82 kilohms) too large—check the value and replace, if necessary. 3. Check for same possible solutions as shown for "No LED Indication on Low Position."

Continuity/Voltage Tester

Continuity/Voltage Tester Electrical Specifications

DESCRIPTION: Equipped with built-in test leads with convenient test probe and alligator clip. Capable of locating electrical continuity of cable assemblies or any electrical conductor. Limited use as resistance indicator and semiconductor diode and transistor tester. Capable of measuring up to 200 volts dc and ac (rms) on high voltage range. Diode protected.

```
MODEL: E64A-1
• DIMENSIONS: 3.75" × 2.5" × 2.2"
• INTERNAL SOURCE: 9-volts/dc alkaline battery
• CONTINUITY FUNCTION:
    Maximum Resistance Value: _____ ohms
• VOLTAGE FUNCTION:
    Low Range (Minimum Value): _____ volts
    Low Range (Maximum Value): 50 volts dc
    High Range (Minimum Value): _____ volts
    High Range (Maximum Value): approximately 200 volts dc
• AC VOLTAGE MEASUREMENTS:
    RMS Values: same as dc specifications
    Peak Value: approximately 285 volts
• TEST LEAD VOLTAGE INSULATION: 2 kilovolts
```

FIGURE 4-59 Sample spec sheet

The final phase of this project involves determining the important electrical specifications for the tester. The format of the *continuity/voltage tester* specification sheet shown in figure 4-59 is very common to many test equipment specification sheets developed by manufacturers of electronic test equipment. The values that must be filled in on this spec sheet can be determined by using a VTVM or VOM, a resistance decade box, a low-voltage dc source, and high-voltage dc and ac sources. The following procedures are designed to determine only the minimum and maximum values of resistance and voltage. Intermediate values are not included in the electrical specifications.

Continuity Function. The *maximum* resistance value that will produce a detectable LED indication will be determined in this test procedure.

Connect the output leads of the tester to the output terminals of the resistance decade box as illustrated in figure 4-60. Rotate the rotary switches of the resistance decade box to a point where the LED indication is no longer detectable. Record the value of resistance on the decade box in the line, "Maximum Resistance Value," on the spec sheet.

Low-Voltage Function. The low-voltage function test determines the *minimum* and *maximum* detectable voltage values in the low voltage range. Connect your tester as illustrated in figure 4-61.

Slowly increase the output voltage from the low-voltage source while monitoring the VTVM and LED. On the spec sheet, record the minimum dc voltage required to give a detectable LED indication.

Increase the output dc voltage from the supply to approximately 50 volts dc and observe the brilliance of the LED. This is the maximum voltage that can be applied in the low range.

High-Voltage Function. The high-voltage function test determines the *minimum* and *maximum* voltages detectable in the high-voltage range. Connect your tester as illustrated in figure 4-62.

Slowly increase the output voltage from the high-voltage source. On the spec sheet, record the voltage, indicated on the VTVM, that is required to ignite the neon lamp. This value represents the minimum voltage detectable in the high-voltage range.

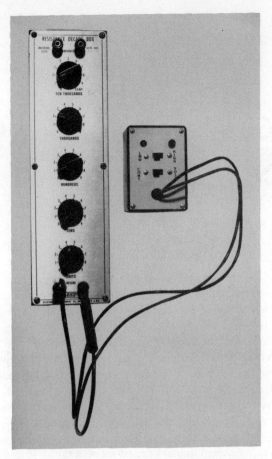

FIGURE 4-60 Test set-up for maximum resistance determination

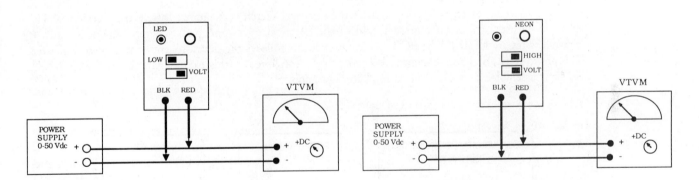

FIGURE 4-61 Test set-up for low-voltage specification

FIGURE 4-62 Test set-up for high-voltage specification

Increase the voltage to approximately 200 volts dc and observe the brilliance of the neon lamp. This is the safe, maximum detectable voltage in the high-voltage range.

Ac Voltage Specifications. The ac voltage specifications indicated on the spec sheet are given in *RMS* and *peak* values. The RMS value is equal to the minimum and maximum dc values for both the low and the high ranges. The maximum peak value, however, is approximately 285 volts.

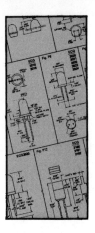

Project 5: Blinking LED Circuit

The blinking LED circuit consists of two bipolar transistors connected as an astable multivibrator. Two pairs of series-connected miniature LEDs function as the indicating devices as well as the circuit load components. The frequency of the flashing rates can be adjusted by changing the values of the RC bias.

This simple circuit will be used to introduce a second method of printed-circuit-board design and fabrication. The technique introduced in Project 2 is a relatively simple but effective method of producing a low-density, printed-circuit board. This method, however, is limited because only one board can be made by this process. If a second board of the same circuit is needed, the entire tape-up procedure must be repeated. This direct tape-up method, therefore, is ideal for a one-of-a-kind circuit board, but it is not recommended if several boards of the same circuit are needed.

The stages of this project are illustrated in figure 4-63. Beginning with the circuit breadboarding exercise, each student is required to build the circuit on a simple springboard and make the necessary circuit modifications. The PCB planning exercise, which follows, is one of the more important steps in the process. The determination of where the components will be placed and the electrical sketch are completed in this exercise.

The artwork master is developed after the electrical sketch is examined for electrical accuracy. The artwork consists of a taped representation of the electrical traces and pads where component leads or external wires are located. The PCB fabrication process follows the development of the artwork master. The photo-resist, or photo-chemical, process will be employed in this exercise. In the final exercise the circuit is assembled and tested in its final configuration.

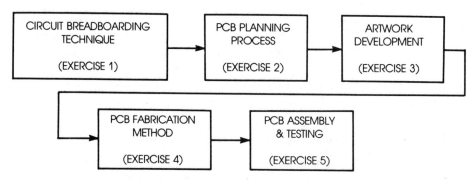

FIGURE 4-63 Block diagram of Project 2 construction sequence

188   Prototype Construction Projects

EXERCISE 1: CIRCUIT BREADBOARDING TECHNIQUE

Purpose

- To identify all components represented on the schematic diagram.
- To use the breadboarding technique to examine the action of the circuit.
- To perform circuit adjustments to the breadboarded circuit and to document changes in the component values.

Components & Equipment

- Component kit of blinking LED circuit
- Springboard with insertion tool

Procedure

Identify each symbol on the schematic diagram of the blinking LED circuit (figure 4-64). Note that the actual component values are not included on the schematic diagram.

Remove all of the components from the parts kit and examine each part. Following is a list of the components identified by their appropriate reference designations and the acceptable range of values for proper circuit operation.

COMPONENT	RANGE OF VALUE
R1 and R4	150–300 ohms, ¼-watt carbon
R2 and R3	22–39 kilohms, ¼-watt carbon
C1 and C2	10–50 microfarads, 10 volts dc or greater
Q1 and Q2	Npn silicon transistor (2N3904, 2N4400, 2N2222, etc.)
LED 1–4	Miniature LED (XC526 or equivalent), red, yellow, green, or clear
B1	9-volt dc transistor battery
Miscellaneous: 9-volt battery snap	

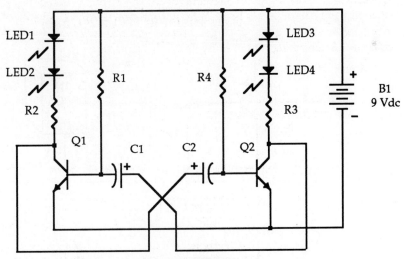

FIGURE 4-64 Schematic diagram of blinking LED circuit

Blinking LED Circuit

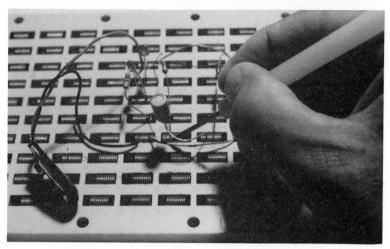

FIGURE 4-65 Breadboarding with springboard system

Write the *exact* value of each component in the kit directly on the schematic diagram beside the appropriate symbol or reference designation (not required for the LEDs).

The springboard system of circuit breadboarding is an efficient and effective means of constructing simple to moderately complex circuits. Components are fastened directly to springs with the pointed insertion tool as illustrated in figure 4-65.

Construct the blinking LED circuit on the springboard. Check the following elements before applying power to the circuit:

1. Polarities of electrolytic capacitors.
2. Transistor elements—emitter, base, and collectors.
3. LED elements—anode and cathode.

The lead elements for transistors and LEDs are identified in figure 4-66.

Apply voltage to the circuit with a power supply or a 9-volt battery. Be careful to connect the polarities properly when connecting the voltage source to the circuit.

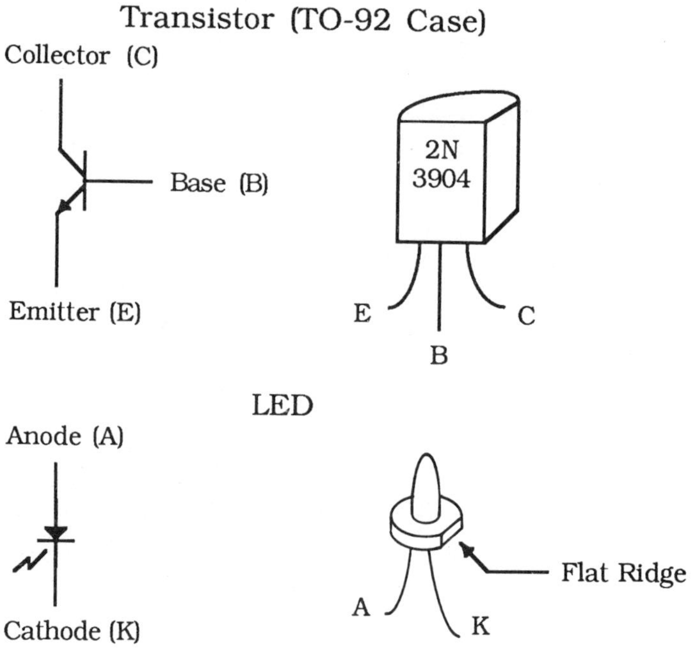

FIGURE 4-66 Transistor and LED lead identification

190 Prototype Construction Projects

Observe the circuit action, if any. A pair of LEDs will flash alternately at a constant interval of time to indicate proper circuit operation. If the circuit does not function properly when power is applied, check for the following conditions:

1. Polarity of the voltage source
2. Polarity of the electrolytic capacitors
3. Incorrect resistor values
4. Incorrect arrangement of transistor leads—base, emitter, and collector
5. Incorrect arrangement of LED leads

The flashing interval may be adjusted by changing the values of resistors R2 and R3 or capacitors C1 and C2. Increasing the values of these components will increase the flashing interval and decreasing the values will decrease the flashing interval.

It is recommended that only one component change be made initially. Either the set of resistors or the set of capacitors can be changed. A combination of value substitutions may then be attempted to examine the resultant circuit operation. Document the final component value directly on the schematic diagram beside the appropriate symbol.

EXERCISE 2: PCB PLANNING PROCESS

Purpose

- To plan the physical layout of the components to be mounted on the PCB.
- To provide an electrical sketch or trace pattern around the physical layout diagram.

Components & Equipment

- Selection of layout problems
- Graph paper (10 squares/inch)
- Light table

Procedure

The PCB will be designed around a picture of a vehicle (see figure 4-67). A careful examination of the picture reveals four areas on the vehicle marked by dots. These dots correspond to the exact location of the four blinking LEDs.

The PCB design problem, therefore, is centered around the predetermined LED locations. The other circuit components must be located around the LED positions.

FIGURE 4-67 Location of LEDs

The following requirements must be considered before any attempt is made to design the printed-circuit board for this circuit:

1. All of the components will be mounted and soldered on the *foil side* of the circuit board (see figure 4-68).
2. Only the LED lense will be exposed at each designated location on the vehicle (as in figure 4-67).
3. All of the components and traces must be located *inside* of the vehicle proper (see figure 4-69).
4. All radial-lead and axial-lead components must be vertical or horizontal to the outside borders of the picture.
5. All transistors and LED leads must conform to the patterns illustrated in figure 4-70.
6. The artwork master (tape-up) must be developed as viewed from the foil side.
7. The battery will be fastened directly to the foil side of the circuit board.

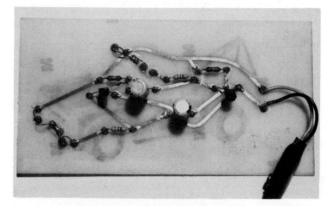

FIGURE 4-68 Mounting of components

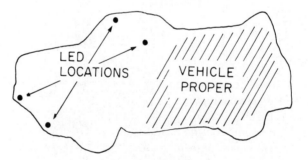

FIGURE 4-69 Location of components and traces

FIGURE 4-70 Arrangement of component leads

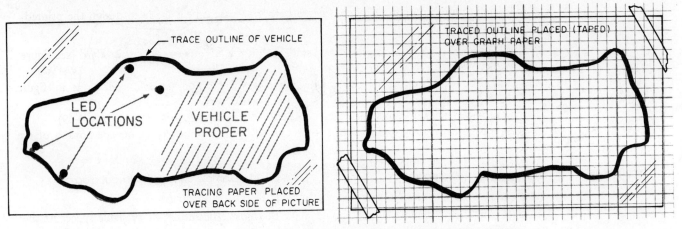

FIGURE 4-71 Transfer of vehicle outline and LED positions

To begin the PCB planning process, trace the outline of the vehicle and the LED locations and tape the traced drawing to a sheet of graph paper (see figure 4-71).

All axial and radial component leads should now be formed as illustrated in figure 4-72. The distance between leads will determine the placement of donut pads during the layout procedure which follows.

Before the electrical sketch is attempted, the physical location of each component with respect to the LED locations must be determined. All sketch work should be done on the graph paper prepared in the preceding step. One method of determining component placement is to begin by positioning the components as they are represented on the schematic diagram (see figure 4-73).

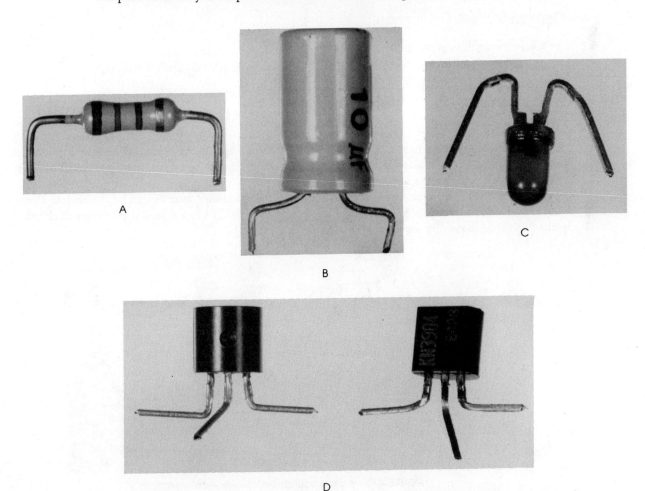

FIGURE 4-72 Lead preparation for each component: (A) axial components, (B) radial components, (C) LEDs, (D) transistors

Blinking LED Circuit 193

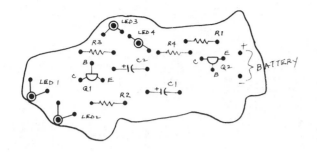

FIGURE 4-73 Component layout sketch preparation

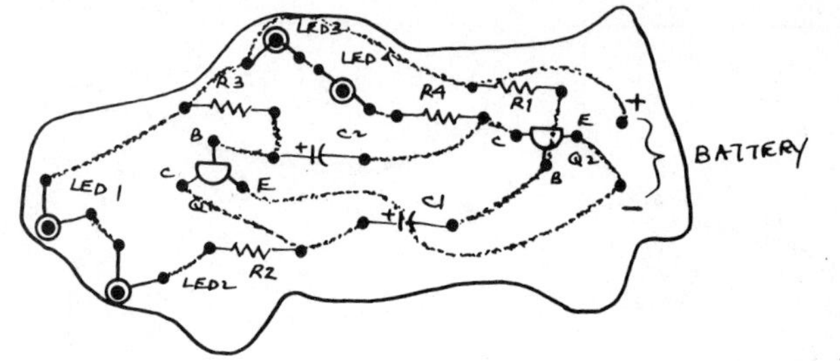

FIGURE 4-74 Example component layout diagram

The ideal component layout diagram should make the most efficient use of available space. At the same time, passive components (resistors, capacitors, chokes, etc.) should be uniformly distributed around the active components (transistors, LEDs, integrated circuits). An example layout diagram is illustrated in figure 4-74.

EXERCISE 3: ARTWORK DEVELOPMENT

Purpose

- To produce the 1:1 positive artwork master.

Components & Equipment

- Clear acetate film (sheet protectors)
- PC board drafting aids: ⅛" (outside diameter) donut pads, 3/16" black layout tape
- X-acto knife or single-edged razor blade

Procedure

The electrical sketch which was developed on the graph paper in the preceding exercise will serve as the basis of producing the artwork master. Tape a piece of clear acetate film (sheet protector) over the electrical sketch as illustrated in figure 4-75.

The following procedure involves applying donut pads and tape as indicated on the sketch. Donut pads are placed at all component lead or external wire locations. The black tape is applied over all electrical paths (traces) as shown on the sketch.

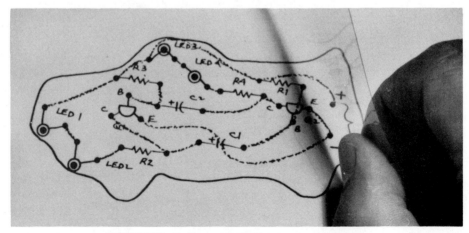

FIGURE 4-75 Taping acetate film on electrical sketch

194 Prototype Construction Projects

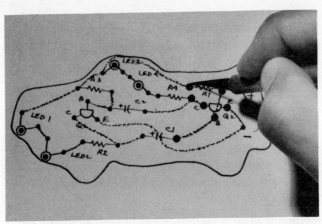
FIGURE 4-76 Applying donut pads to lead locations

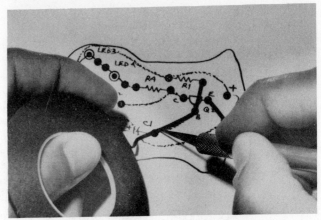
FIGURE 4-77 Applying black tape for trace runs

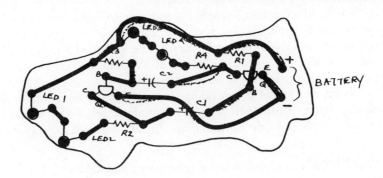
FIGURE 4-78 Completed artwork master

Before performing this step, review the recommended practices for trace runs in Table 4-4. These practices should be observed when developing the artwork master.

The example sketch is used to illustrate the steps for producing the artwork master for this project (see figures 4-76 through 4-78).

EXERCISE 4: PCB FABRICATION METHOD

Purpose

- To produce a single-sided PC board using the photochemical process.
- To introduce one system of producing PC boards by the photochemical process and to compare this system with other techniques.

Components & Equipment

- Single-sided, blank PC board material
- Wet-dry, abrasive paper (250 to 350 grade)
- Positive photoresist
- Photoresist developer
- Ultraviolet exposure system
- Ammonium persulfate crystals (etchant)
- Diluted liquid soldering flux
- Laboratory oven
- Etching machine

Blinking LED Circuit   195

Procedure

Cleaning the Copper-Clad Board. The copper surface of the board must be thoroughly cleaned for the photoresist to have good adhesion. In most cases, water with household cleanser is sufficient for cleaning the copper surface; however, if the copper surface is highly corroded, other cleaning methods must be employed. The use of extra-fine, wet-dry, abrasive paper under running tap water will remove surface grease and corrosion and will provide a smooth, clean surface. A circular motion is recommended for this cleaning method.

Drying the Cleaned Board. The cleaned PC board can be dried at room temperature or in an oven. If an oven is used, it should be set at approximately 110 to 125 degrees Fahrenheit. This temperature range is also the optimum range for the resist baking step. Before proceeding further, be sure that the copper surface is completely dry and is not contaminated by fingerprints.

Applying Photoresist to the PC Board. The photoresist solution used to coat the copper-foil surface is extremely sensitive to high-intensity, white light; therefore, the use of subdued white or yellow light is recommended for a safe application condition.

The liquid resist must be applied in such a manner as to produce a thin, consistent coating. The resist material may be applied by a number of acceptable methods. An acid brush may be used to apply an initial layer of the resist to one end of PC board. The PC board is then held in a vertical position so the excess resist can easily flow down to the lower end. When the resist reaches the lower end, the board is laid flat to insure an even distribution of the resist material.

Another technique of providing an even distribution of the resist material requires the use of a spinning source. A converted phonograph turntable, set at 45 rpm, can be used as a means of distributing the resist evenly about the copper surface. An air brush used to spray the resist will produce a high-quality, coated PC board.

CAUTION: Most resist materials are extremely flammable and should not be sprayed near open flames.

The dip coating method can also be used to apply the resist material to the copper surface. Commercial PC board, dip coating equipment is available for purchase at moderate prices. This method, however, is normally used for large boards or for the coating of double-sided PC boards.

Baking the Coated PC Board. The coated PC board must be allowed to dry thoroughly before proceeding any further. The required drying or baking time depends on the thickness of the coating.

A laboratory oven is used to bake the coated PC board. The oven temperature is critical; it should not exceed 125 degrees Fahrenheit. A long baking time at a low temperature is preferred over a short time at a high temperature.

Cooling the Baked PC Board. After the photoresist is baked onto the copper surface, it must be cooled "to the touch." This step is important to insure that the resist is dry and not tacky as may be its condition when hot.

Exposing the Photosensitized PC Board. Ultraviolet light is the primary exposing source, but other high intensity light sources may also be used. The exposure time depends on the type of resist material, the thickness of the coated board, and the type of light source. A little experimentation is the best method to determine correct exposure times.

The positive artwork master is placed over the photosensitized copper surface. The artwork master must be placed flat and firmly against the copper surface to avoid exposing under the trace or pad areas.

*NOTE: Although the artwork master is always placed **between** the light source and the photosensitized copper, it can be easily exposed through the "wrong" side. Exercise caution and be certain that the correct side of the artwork master is exposed before proceeding.*

Developing the Exposed PC Board. The developing process dissolves, or removes, the photoresist only at the exposed areas of the PC board. The exposed board is immersed in a *glass* tray that contains about 1/8 inch of photoresist developing solution. The tray can be agitated by lifting one end and then the other during the developing period. The developing time is dependent on the thickness of the resist coating, the exposure time, and the type of resist material. A properly developed board will be very obvious. The pads and traces will be well defined with the exposed areas free of any resist "cloud."

CAUTION: Most photoresist developers consist of a strong alkaline base (e.g. household lye). Rubber gloves and/or wooden tweezers should be used when handling the PC board in the solution.

Rinsing and Baking the Developed PC Board. The developed PC board must be rinsed and baked again in the oven. This final baking insures that the resist areas that may have been softened by the developing process are solidified again. The baking time is approximately 10 to 15 minutes at 110 to 125 degrees Fahrenheit.

Etching the PC Board. In the etching process, the unwanted copper is removed to achieve the identical artwork master trace patterns. The process involves the use of an etching chemical and an etching machine. The selection of the etching solution depends on the resist material used to transfer the image to the copper-clad board. The most widely used types of etching solutions are:

1. Ferric chloride: This etchant is widely accepted and used in industry as an acidic etching solution. It is relatively inexpensive and is able to absorb large amounts of dissolved copper while maintaining its etching ability.
2. Ammonium persulfate: This type of etching chemical is the most commonly used alkaline-based etching solution. It has a comparatively slow etching rate and a low dissolved-copper capacity, but it is comparatively easy to handle and it provides less of a disposal problem than the acid-based etchants.
3. Cupric chloride: This acid-based solution etches at about half the rate of time as ferric chloride. Its ability to dissolve copper is comparatively low, but it can be easily regenerated by adding hydrochloric acid.
4. Chromic-sulfuric acids: Mixtures of chromic-sulfuric acids are not particularly effective as copper etchants. They are normally used on solder-plated PC boards. Mixtures made from chromic-sulfuric acids have a long life, but they are difficult to handle and to dispose of.

Commercial PC board etching machines can be purchased or fabricated. The etching machine must contain a heating source and a means of applying the etchant to the PC board. The most common types of etching machines are based on *immersion*, *bubble*, and *spray* principles. The machines developed from these principles are very simple, but they must be constructed from materials that will not chemically alter the etching solution.

Immersion etching involves immersing the developed PC board in a container of etchant. The process is relatively slow, but it can be increased by providing a heating source and by agitation. A simple container can be constructed from an old car battery casing after the plates and the electrolyte solution have been removed. An aquarium immersion heater can be used as the heating source.

In bubble etching, air is introduced into the immersion etching system as the agitating source. A simple bubble etcher can be constructed by equipping the battery-casing immersion system with an aquarium air pump to produce the bubbles.

In spray etching, the PC boards are held by small brackets and spray nozzles are moved back and forth to apply an even stream of etchant spray to the boards.

Removing Resist from the Foil Pattern and Flux Coating. This is the final step in the PC board fabrication process. After the etching process, the foil pattern is still coated with resist material. Because this resist material may prevent proper solder flow, it must be removed. The wet-dry abrasive paper used in the cleaning process is utilized in this step to remove the resist material.

When the board is dry, a thin layer of diluted-rosin soldering flux is applied to the foil pattern. This application of rosin flux will prevent oxidation or corrosion from forming on the copper-foil pattern.

After the resist has been removed and the soldering flux has been applied to the foil pattern, the four LED lense positions are drilled. This operation will be performed in the following exercise.

Following is a review of the printed-circuit-board fabrication procedure:

1. Clean the copper-clad board.

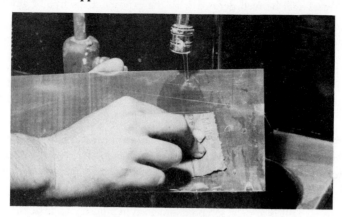

2. Dry the cleaned board.

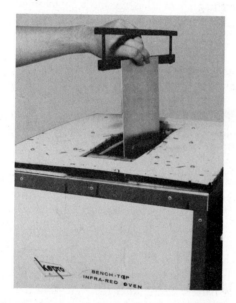

3. Apply photoresist to the PC board.

4. Bake the coated PC board. Repeat step 2 using the same equipment.
5. Cool the baked PC board. The baked board should be cooled in a dark, dust-free enclosure.
6. Expose the photosensitized PC board.

7. Develop the exposed PC board.

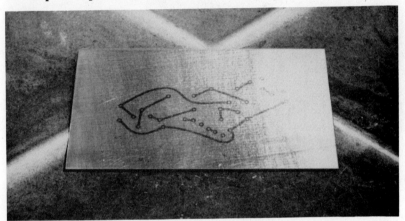

8. Rinse and bake the developed PC board. Rinse the developed board with tap water and bake it in the same equipment used in step 4. Be careful not to scratch the patterns during the rinsing and baking procedures.
9. Etch the PC board.

Blinking LED Circuit 199

10. Remove the resist from the foil pattern and the flux coating. Resist can be removed with wet-dry abrasive paper as in step 1.

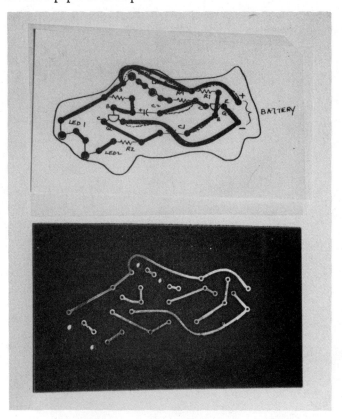

EXERCISE 5: PCB ASSEMBLY AND TESTING

Purpose

- To drill the necessary holes for the four LED locations.
- To apply the picture of the vehicle to the etched PC board.
- To solder each component to the PC board.

Components & Equipment

- Etched PC board
- Picture of vehicle
- Circuit components
- Spray adhesive
- Soldering station, solder, solder stand
- Drill press and twist drill set

Procedure

The four LED locations must be drilled to the proper size so that only the lense portion is exposed through the circuit board. The picture of the vehicle is then applied to the circuit board so that the LED locations on the picture are aligned with the holes on the PC board. Spray adhesive is applied *to the PC board* and the picture is placed carefully over the four LED holes.

FIGURE 4-79 Completed project

The LED holes on the picture are then enlarged to the size of the holes drilled on the PC board and the PC board is trimmed to its final dimension with the squaring shear.

Finally, the PC board is assembled, using the electrical sketch to determine the correct locations of all components to be soldered to the PC board. The completed project is shown in figure 4-79.

Project 6: Adjustable, Bipolar, Regulated Power Supply

An adjustable, bipolar, regulated power supply capable of supplying approximately ±15 volts dc at 1 amp will be constructed in this advanced prototype construction project. Unlike the first two projects, the electronic circuitry in this project is relatively complex and requires more involved planning in the areas of chassis fabrication, PC board design, and chassis assembly techniques. The block diagram in figure 4-80 illustrates the construction stages of this project.

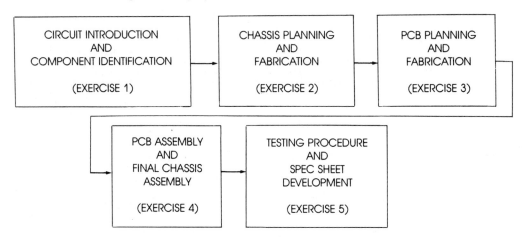

FIGURE 4-80 Block diagram of construction sequence

The instructions provided in the exercises for this project are not as detailed as those presented in projects 1 and 2. The exercise on chassis planning and fabrication includes only a suggested sheet metal layout diagram for a chassis that is fabricated from a flat metal stock. Commercially available chassis can be purchased and are also considered in the initial planning stages of this project. Reference texts that contain more detailed instruction on the various sheet metal processes should be consulted throughout this project.

The exercise on PCB planning and fabrication includes suggestions and special techniques to assist in the PC board layout process. The PC board fabrication technique introduced in Project 3 is used in this project; that is, development of a 1:1 positive artwork master that is used in the photochemical method of fabricating PC boards. A second method of fabricating PC boards without chemicals is also introduced in this unit. In this innovative system of producing a PC board without chemicals, the laminated copper board is routed to produce the insulated areas between the pads and adjacent traces.

The chassis assembly and wiring techniques introduced in the Projects 2 and 3 should be reviewed before beginning Exercise 4: PCB Assembly and Final Chassis Assembly. In this exercise, the PC board and the external components that are mounted directly on the chassis are connected according to the schematic and a wiring diagram.

In the final exercise, entitled Testing Procedure and Spec Sheet Development, the procedures for turning the power supply on and for troubleshooting the power supply are covered. The power supply specifications are also developed in this exercise.

EXERCISE 1: CIRCUIT INTRODUCTION AND COMPONENT IDENTIFICATION

Circuit Introduction

The schematic diagram of the adjustable, bipolar, regulated power supply (figure 4-81) is used exclusively to introduce the circuit components and to explain basic circuit action. A three-prong AC plug and line cord connects the power supply primary circuit to the 117-volt ac (sine wave) line voltage. The primary circuit consists of the line cord, the fuse (F1), the switch (SW1), and the primary winding of transformer (T1).

The 117-volt ac line voltage is stepped down by transformer action to approximately 25 volts ac across the full secondary winding of T1. The secondary winding of T1, which is center-tapped, serves as the *circuit ground* reference. All voltage measurements around the test points of this circuit are referenced to this point.

All regulated power supplies contain an unregulated, or bulk, supply section that feeds the voltage regulator section. The unregulated supply section consists of the four diodes (D1–D4) and capacitors (C1–C4). The function of the diode network is to rectify, or to change the stepped-down, sine-wave voltage to a pulsating dc voltage. When the voltage at the junction of D3 and D4 is measured with respect to ground, it should be a positive dc voltage. On the other hand, the voltage measured at the junction of D1 and D2, with respect to ground, should be a negative dc voltage. Capacitors C1 through C4 serve as filters to the low- and high-ripple frequency variations from the diode rectifier network.

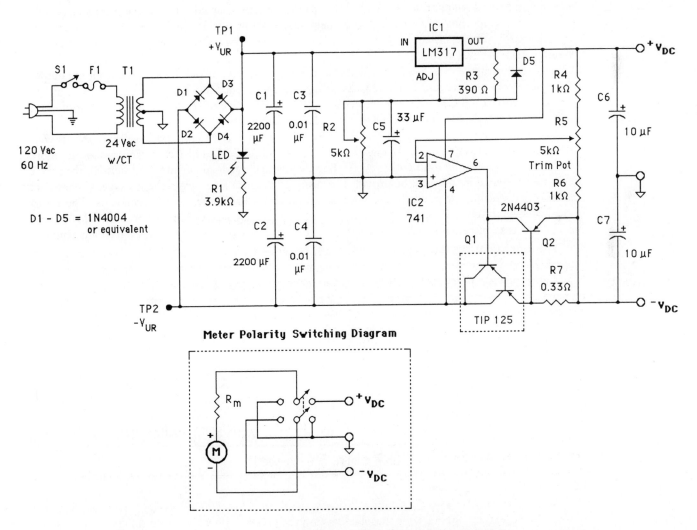

FIGURE 4-81 Adjustable bipolar power supply

Adjustable, Bipolar, Regulated Power Supply

The series combination of R1 and LED1 acts as the indicator lamp on the front panel of the supply.

The function of the regulator section is to maintain a constant voltage at any preset value regardless of any load current changes within the design specifications. The basic regulator section consists of a positive, adjustable, regulator chip, IC1, with its supporting components, resistor R3, diode D5, capacitor C5, and potentiometer R2. IC1 is an LM317 regulator in a three-terminal, plastic package. The three terminals are labeled input (IN), output (OUT), and adjust (ADJ). A load connected to the positive output and the ground terminals is in series with the input-to-output terminals of IC1. For this reason, IC1 is classified as a *series* regulator. The positive output of the power supply can be adjusted with R2 from approximately 1.2 to 15 volts dc.

The regulator chip is designed to maintain a constant set voltage even if changing load currents cause the unregulated supply voltage to fall and the ripple voltage to increase. Voltage variations occurring at the input of the regulator are absorbed across the input-to-output terminals of IC1. The only limitation of this regulator chip is that the input-to-output voltage differential must be greater than 2 volts and less than 40 volts. The LM317 chip also provides three unique internal protection circuits to prevent any possible fault condition from damaging the regulator. The first internal protection circuit protects the chip from being internally fused under short-circuit load conditions. This protection feature is known as *current limiting*. The second protection circuit, known as the *safe-area protection*, decreases the output current demand as the input-to-output differential voltage is increased. The final protection circuit, known as the *thermal-overload*, shuts down the regulator when the internal chip temperature rises above approximately 200 degrees Fahrenheit.

The negative regulator section consists of a 741 op-amp (operational amplifier), a chip (IC2), and a *Darlington-pair* transistor package (Q1) which is a pair of directly coupled transistor amplifiers in a single package. Other supporting components for the negative regulator section consist of resistors R4 and R6 and trim pot R5. The function of R5 is to maintain a balance between the positive and negative output voltages with respect to ground.

A meter polarity switch (SW2) is included at the output section of the supply to monitor the output voltage across the positive and negative jacks with respect to ground.

Capacitors C6 and C7, which are located at the output, improve the performance of the positive and negative regulators by preventing oscillation under all load conditions.

The power-dissipating capacity of the positive regulator (LM317) package and the power, Darlington-pair transistor (TIP125) must be considered. Since the LM317 chip cannot dissipate more than 22 watts of power by itself without activating the thermal-overload circuitry and shutting the circuit down, some external means must be provided to dissipate the heat away from the package. Commercial heat sinks are available to remove the excessive heat from the package to allow larger amounts of power to be dissipated by the chip. A simple but effective means of providing heat sinking for both the LM317 and the TIP125 is to place the metal tab in contact with a heat sink constructed of 1/16-inch aluminum of approximately 25 square inches.

WORKSHEET 4-7: Component Identification and Inventory Check

Locate each component listed below in the parts kit and check it off.

CHECK (✔)	REFERENCE DESIGNATION	DESCRIPTION
( )	R1	3.9 kilohm, 5%, ½ watt
( )	R2	5-kilohm linear pot with SPST switch (SW1)
( )	R3	390 ohm, 5%, ½ watt

CHECK (✔)	REFERENCE DESIGNATION	DESCRIPTION
	R4 & R6	1 kilohm, 5%, ½ watt
	R5	5 kilohm trim pot
	R7	0.33 ohm, ½ watt
	R_M	Value to be determined
	C1 & C2	2200 µF, 50-volts dc, electrolytic
	C3 & C4	0.01 µF, 100-volts dc, ceramic
	C5	33 µF, 50-volts dc, electrolytic
	C6 & C7	10 µF, 50-volts dc, electrolytic
	T1	24-volts ac with center-tapped 2-amp transformer
	D1–D5	1N4004, 400-volts dc rectifier diodes
	LED	Red jumbo LED (T-1 case)
	Q1	T1P125, pnp darlington pair transistor
	Q2	2N4403 pnp silicon switching transistor
	IC1	LM317 positive voltage regulator (TO-220 case)
	IC2	741 operational amplifier (8-pin dip package)
	F1	1-amp, 250-volt fuse
	S2	DPDT slide switch
	M	50 µA, ≈ 2 kilohm internal resistance
		3-prong line cord
		Chassis mount fuse holder
		LED panel mount clip with retainer ring
		8-pin dip socket
		Red, black, and green binding posts
		Miscellaneous hardware (knobs, machine screws, nuts, washers)

EXERCISE 2: CHASSIS PLANNING AND FABRICATION

The chassis for the power supply can be of several configurations. The common commercially available chassis configurations are illustrated in figure 4-82. Any of the chassis styles illustrated in this figure can be used for the power supply in this project, instead of designing and fabricating a custom chassis and enclosure.

Any prototype fabrication shop equipped with the proper sheet metal equipment can produce a custom chassis. If a shop is not well equipped, however, only basic hand sheet metal tools need be used. These basic hand tools include those that perform the operations of drilling, punching, filing, and shearing or nibbling.

The chassis suggested for this power supply consists of three basic sections: the *front panel*, the *chassis frame*, and the *deck plate*. If an *open*-box style, prefabricated chassis is used, only a minimum amount of sheet metal layout and sheet metal processes are required to produce the chassis unit. The assembly drawing in figure 4-83 illustrates how the sections are fastened to each other to form the chassis unit. The layout drawings (figures 4-84 through 4-86) are provided for suggested front panel, frame, and deck plate layouts of the items that will be mounted directly on these sections. The layout drawings of the frame and deck plate also show the hole locations of the items that will be fastened to these sections.

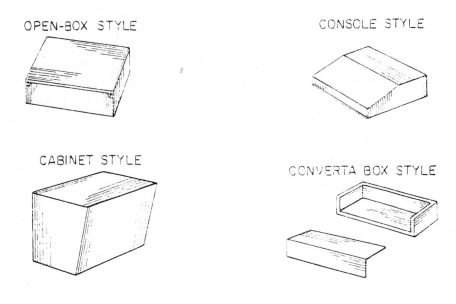

FIGURE 4-82 Common chassis styles

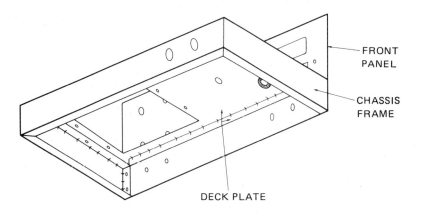

FIGURE 4-83 Chassis assembly drawing

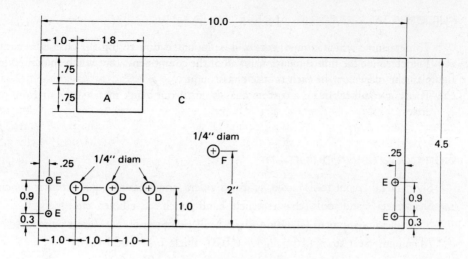

FIGURE 4-84 Front panel layout

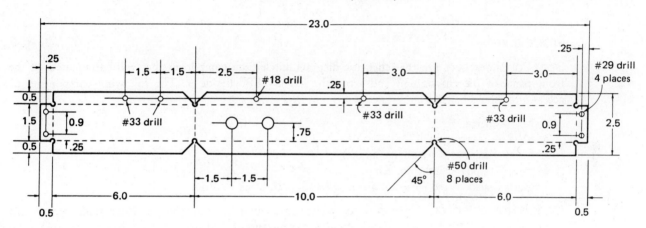

FIGURE 4-85 Chassis frame

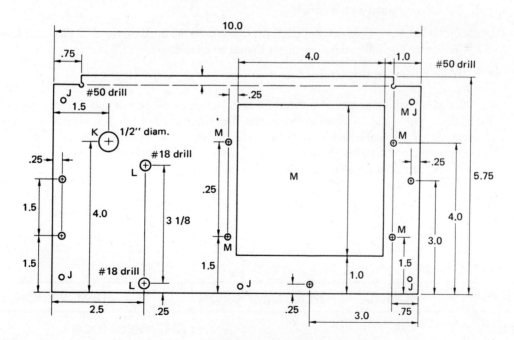

FIGURE 4-86 Deck plate

Adjustable, Bipolar, Regulated Power Supply 207

Purpose

- To determine which components will be mounted directly on the chassis.
- To determine the hole dimensions of all of the components that will be mounted on the chassis and to plan the placement of each on the chassis unit.
- To design and fabricate a custom chassis unit from blank metal stock or from prefabricated chassis boxes.

Components & Equipment

- Sheet metal layout tools (scale, scribe, dividers, combination square, center punch, etc.)
- Sheet metal hand tools (chassis punch, hand punch set, nibbler, etc.)
- Sheet metal equipment (squaring shear, brake, corner notcher, turret punch, etc.)
- Aluminum sheet stock (16 gauge = 0.050" thick, type 3003-H14)
- Optional: prefab chassis box (approximately 6" × 10" × 1.5")

Procedure

Front Panel Section. On the schematic diagram, circle each component that will be mounted on the front panel of the chassis unit and write the letters "FP" beside it. The following components or accessories should be circled:

1. M1 = meter
2. SW2 = rotary switch
3. R2 = 5-kilohm pot (voltage adjust)
4. LED1 = power indicator lamp
5. Output jacks = red (+Vdc), green (ground), and black (−Vdc)

Chassis Frame Section. Circle each component that will be mounted on the chassis frame section and write the letters "CF" beside it. The following components should be circled:

1. The 3-prong line cord
2. The chassis-mount fuse holder

Deck Plate Section. Circle each component that will be mounted on the deck plate and write the letters "DP" beside it. The following component should be circled:

1. T1 = power transformer

A chassis layout diagram is provided for each section of the chassis unit (figure 4-84 through 4-86). Each hole in each section is identified by a letter. The function of each hole is identified in Worksheet 4-8. The purpose of this worksheet is not only to identify each hole on the chassis, but also to determine the *actual* dimensions of the item that will occupy the assigned hole. This requires that the mounting dimension of each item be measured. Write the dimensions in the appropriate space on the worksheet and transfer them directly to the layout diagrams.

WORKSHEET 4-8: Chassis Unit Hole Assignment

SECTION	HOLE ID	HOLE DIMENSION	DESCRIPTION
FRONT PANEL	A		M1, meter face
	B		R2, 5 kilohm pot (voltage adjust)

SECTION	HOLE ID	HOLE DIMENSION	DESCRIPTION
	C		SW2, meter function switch
	D		Red, green, and black output jacks
	E		4 front panel mounting screws (machine screw 4-40 × ¼")
	F		LED1 mounting clip
CHASSIS FRAME	G		4 front panel mounting screws
	H		3-prong line cord
	I		Chassis-mount fuse holder
DECK PLATE	J		Deck plate mounting holes to chassis frame (machine screws 4-40 × ¼")
	K		Rubber grommet
	L		T1, power transformer mounting holes (machine screws 6-32 × ½")
	M		PC board mounting screws (#4 × ¼" sheet metal screws)
	N		PC board

EXERCISE 3: PCB PLANNING AND FABRICATION

The PC board for the power supply will be mounted directly over the large square hole on the deck plate. The final size of the PC board, therefore, is set before the component layout of the circuit is considered. This size limitation could become an obstacle during the design process, but the allotted 4" square hole is more than generous and should encourage conservative layout planning.

There are several other PC board design limitations. These limitations are included to introduce design techniques or to create obstacles for the designer. A list of design limitations is provided in the procedure section of this exercise.

The photochemical method of PC board fabrication was utilized in Project 5. This method of producing PC boards has several advantages. It has the ability to produce more than one board of the same circuit with good results. By maintaining a system of monitoring the chemicals utilized in this process, it produces high-quality boards with consistent definition. And, it is able to produce high-density circuits more efficiently due the high-resolution aspects of the photographic process.

Despite the advantages of the photochemical method, there are a few drawbacks to this widely used system. Since chemicals are the main ingredients for this process, the importance of educating all students on the proper handling and the hazards of each chemical used cannot be overemphasized. Proper ventilation and an approved eyewash/shower station must be provided in case of any accidental eye or

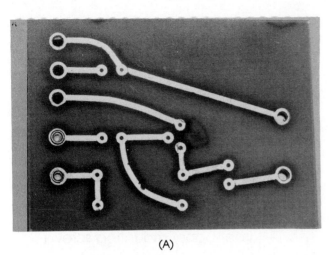 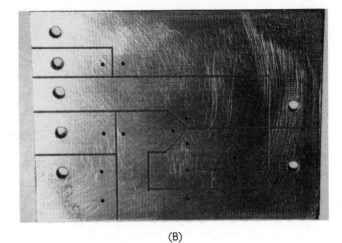

FIGURE 4-87 Comparison of etched and routed PC board: (A) etched PC board, (B) routed PC board

bodily contact with these chemicals. Approved chemical containers or other systems for safe disposal of contaminated or used chemicals must be provided to insure a safe working environment. Finally, the cost of these specialized chemicals will continue to increase and eventually will constitute a major expense item.

A chemical-free system developed by LPKF Pacific Corporation employs a high-speed router bit to produce PC boards by simply routing copper strips from the blank copper-clad board. The operator controls the feed and the depth of cut manually. A comparison of an etched board to a routed PC board of the same circuit is illustrated in figure 4-87.

The routing system is attractive because it is a chemical-free process. This system should also prove to be relatively inexpensive in the long run. A limitation of this system is the difficulty experienced in producing complex, high-density, single-sided or two-sided PC boards. This system, however, is relatively easy to master and requires only an accurate 1:1 layout sketch.

Despite the differences in these two systems of fabricating PC boards, both methods require similar layout planning steps before the actual fabrication is performed.

Purpose

- To develop a physical and electrical layout sketch of the PC board.
- To develop an artwork master for the photochemical or routing process of fabricating the PC board.
- To fabricate the PC board using the photochemical or routing system.

Components & Equipment

- Graph paper (10 squares/inch)
- 1:1 drafting component template
- Component kit
- Artwork material (donut pads, tape, etc.)
- Photochemical PC board equipment
- LPKF 39 System Seeback (LPKF Pacific Corp., San Rafael, CA)

Procedure

PC Board Design Requirements

1. The maximum finished outside dimensions is 5" square. This allows at least 0.5" from the edges for mounting to the deck plate (see figure 4-88).

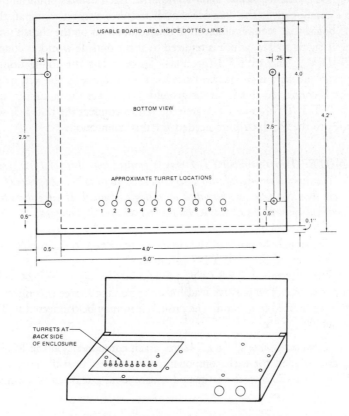

FIGURE 4-88 PCB dimensions and turret locations

2. Each connection from a component mounted on the chassis to a component mounted on the PC board requires a turret terminal. All turret terminals must be placed on *one-edge* of the PC board (see figure 4-89).
3. Both TO-220 package components (IC1 = LM317 and Q1 = TIP125) must be mounted on adequate heat sink and soldered on a separate PC board (see figure 4-90).

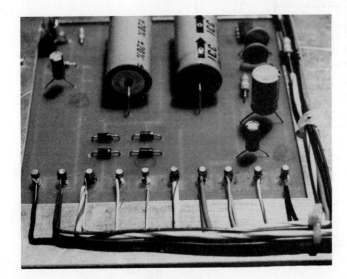

FIGURE 4-89 Turret terminals on PC board

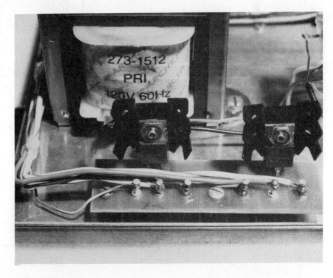

FIGURE 4-90 Heat sink and PC board for IC1 and Q1

Adjustable, Bipolar, Regulated Power Supply

PC Board Design Steps

Determine the number of turrets required. Each *separate* point on the PC board that requires a connection to a component in the "outside world" requires a turret terminal; that is, any electrical connection on the PC board that requires connection to a component on the chassis (see figure 4-91). In this simple example, the ac signal source, which is referred to as the outside world, is connected to the input of the circuit board, TT-1 (V_{in}), and to TT-2 (ground). To complete the connection, two wires are connected to the PC board: one wire is connected from the signal source to the left, or positive (+), side of C1 and a second wire is connected to the circuit ground. Ground is the common connection point of the circuit.

TT-3, TT-4, and TT-5 are required to connect the pot, R3, which is mounted to the chassis. Three wires to the PC board are needed for this connection.

NOTE: *Turret terminal TT-5 may be omitted since this point is a ground, or common point, of the circuit. If this terminal is eliminated, two wires must be soldered to TT-2 (ground). Although this practice is acceptable, no more than two wires are permitted on a single terminal. If more than two wires must be connected to the same electrical point in the circuit, additional turrets must be provided.*

The signal-out point of the circuit is connected to the oscilloscope. This connection is made possible by TT-6 (signal out) and TT-7 (ground).

Turret terminal TT-8 (+V_{cc}) is required to supply the 24 volts dc to the circuit from an external voltage source. The positive lead from the voltage source is connected to this terminal, while the negative lead is connected to ground. The ground lead may be connected to TT-7, the separate ground terminal for the output.

Estimate the area of the PC board. Each component on the PC board takes up space. By determining the area needed for each component, the smallest possible size of the PC board can be estimated. By adding 25%–30% to the estimated square area, a reasonable estimate of actual board size can be made.

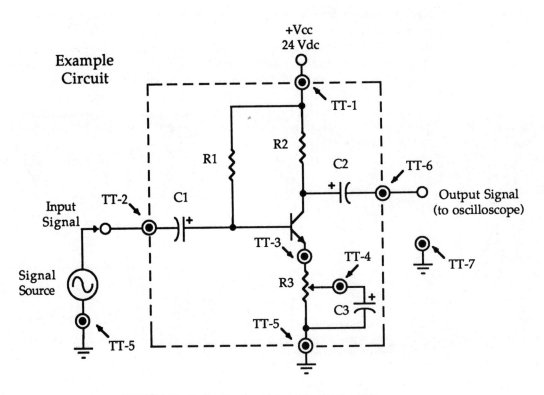

FIGURE 4-91 Determination of turret terminal points

When determining the area to be occupied by a certain component, several factors must be taken into account. First, the component lead, which is fed through a hole from the component side of the board and soldered to a donut-shaped pad on the foil side, occupies space. The pad is an electrical connection; it must not touch any other pads or traces unless their electrical connection is required. Therefore, there must be some clearance around each pad. The area of the pad and the clearance around the pad must be taken into account when estimating the area required by each component.

NOTE: As a rule of thumb, adjacent traces or pads must be at least one-half to one trace width apart.

Another factor to consider when estimating area is that there should be enough room for ease of assembly and for component replacement or PC board repair. If two components are too close to each other, the replacement task could be difficult. Ample room should be provided for reaching in to remove any component with needle nose pliers.

Components that dissipate heat must also be considered. Other components should not be close enough to warm components to be affected by them. Components that normally dissipate heat are resistors and high-power semiconductor devices. The only components in this power supply circuit that may develop some heat are IC1 (LM317) and Q1 (TIP125). These two components, however, are mounted on a separate PC board/heat-sink unit away from the main circuit board.

In Table 4-6, the estimated area for each component in the circuit used in the example is given. An explanation of the methods used to determine the area for each component follows the table.

TABLE 4-6 Area Estimation

COMPONENT	SIZE (L × W)	ESTIMATED AREA (L × W + ALLOWANCES)	TOTAL AREA (SQUARE INCH)
R1	0.38 × 0.13	0.83 × 0.33 = 0.27	0.27
R2, R3	0.25 × 0.09	0.7 × 0.29 = 0.20	0.40
D1	0.13 × 0.06	0.58 × 0.26 = 0.15	0.15
C1	0.5 × 0.1	0.8 × 0.3 = 0.24	0.24
Q1	0.5" diameter	0.5 × 0.5 = 0.25	0.25
3 TURRETS		0.3 × 0.3 = 0.09	0.27
TOTAL			1.58

To estimate the area for axial lead components (components that lay flat on the circuit board; e.g., most fixed resistors, semiconductor diodes, and a number of capacitor types), determine the width of the component and *add* 0.1" or ¼ of the width, whichever is greater, on each side. Determine the length of the component and add 0.1" or ⅙ of the length, whichever is greater, on each end for the leads. Then add the pad diameter to each end to account for the pad and the clearance around it (see figure 4-92.)

The body of a ½-watt carbon resistor measures 0.13" wide and 0.38" long (not including the leads). It will use a ⅛" diameter donut pad.

One-fourth of its width is equal to 0.03". Since 0.1" is greater, it should be added to the actual width *on each side*.

WIDTH: 0.13" + 0.1" + 0.1" = 0.33"

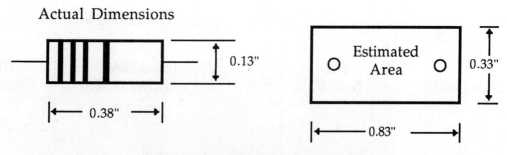

FIGURE 4-92 Estimated area for axial components

Actual Dimensions   Estimated Area

 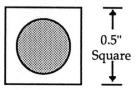

FIGURE 4-93 Estimated area for radial components

One-sixth of its measured length is equal to 0.06". Again, since 0.1" is greater, it should be added to the length on each side. The diameter of the donut pads ($\frac{1}{8}$" = 0.125") should also be added.

LENGTH: 0.38" + 0.1" + 0.1" + 0.125" + 0.125" = 0.83"

The area for this part can then be calculated:

0.33 × 0.83 = 0.27 square inches

The allotted area for radial lead components (components that stand upright because the leads are located on only one end of the component body; e.g., most capacitors, trim pots, and transistors) should be a square area approximately *twice* the measured diameter, but never less than 0.5" square (see figure 4-93 for an illustrated example).

The diameter of a 10 µF, 35-volt, dc capacitor with radial leads measures 0.2". Twice 0.2" is 0.4". Since this is less than the 0.5" minimum square length, a square area of 0.5" × 0.5" is allotted for this component. The estimated square area for this capacitor is

0.5" × 0.5" = 0.25 square inches

All turret terminals must be included in the area estimation. The *minimum* distance between adjacent turrets should be approximately 0.3". This will allow a space of 0.3" × 0.3" square for the required area for each terminal.

Add 30% to 1.58 square inches to get a reasonable area for the completed PC board.

30% of 1.58 = 0.47 square inches

The estimated finished square area of the example circuit is

0.47 + 1.58 = 2.05 square inches

WORKSHEET 4-9: PC Board Area Estimation for Power Supply

Refer to Table 4-6 for an example of an area estimation worksheet. Estimate the area for each component that will be mounted on the PCB.

COMPONENT	SIZE (L × W)	ESTIMATED AREA (L × W + ALLOWANCES)	TOTAL AREA (SQUARE INCH)

COMPONENT	SIZE (L × W)	ESTIMATED AREA (L × W + ALLOWANCES)	TOTAL AREA (SQUARE INCH)
TOTAL SQUARE AREA			
TOTAL SQUARE AREA PLUS 30%			

Prepare the circuit layout sketch. After the area for the PC board is estimated the layout can be prepared. A variety of component layout templates, which assist in the developmental stages of the layout process, are available (see figure 4-94).

The PC layout tools include such basic drafting tools as 4:1–2:1 scale, a $\frac{30}{60}$ and a 45-degree triangle, a drawing pencil, an eraser, a drafting board, and a T-square. Although most components can easily be drawn with these basic tools, templates such as those shown in figure 4-94, which contain most components scaled to size, are almost essential for efficient work.

Cut-outs, or paper dolls, of the components involved in a given layout are also helpful. They can be made from paper or light cardboard and should be labeled with reference designations as on the schematic diagram. They can then be placed or taped on the layout grid paper.

Another useful PC designer's tool is a light table.

The layout procedure normally involves three stages. They include the development of the rough sketch, the component layout diagram, and the positive artwork or trace diagram. To make the rough

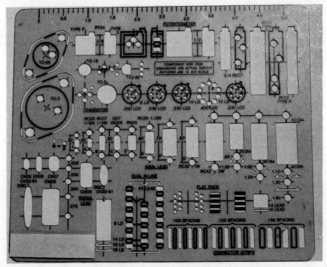

FIGURE 4-94 PC design drafting templates

Adjustable, Bipolar, Regulated Power Supply 215

sketch, take a sheet of 0.1" grid graph paper and mark off the length and width of the PC board (estimated in Worksheet 4-9) at the corners. Arrange the components in essentially the same pattern as those on the schematic so that there are no lead crossovers. Where leads tend to cross, route them under the body of components so that no jumpers are required. After the components are arranged and sketched in place, draw dotted lines to indicate where the leads interconnect. A rough layout sketch for the example circuit is shown in figure 4-95.

To prepare the component layout diagram, remove the completed rough sketch from the drafting board, place it *face down* on the light table, and tape it into place.

NOTE: *Be sure to turn the rough working sketch face down so that the right side of the sketch becomes the left side and the left side becomes the right side. The top and bottom should remain in the same position.*

Tape a sheet of tracing paper on top of the rough sketch and trace the component outlines, including the dots that represent lead locations. Note that the signal flow is from the right to the left of the component layout. The component layout diagram for the example circuit is shown in figure 4-96.

To develop the trace diagram, tape the rough sketch *face up* on the light table. This view is of the foil side of the PC board. If the photochemical PCB process is being used, place a sheet of clear acetate (sheet protector) over the rough sketch and place the appropriate sizes of donut pads or other special artwork patterns over the lead locations. Following the trace pattern, apply PC drafting tape to the sketch. The trace diagram is now ready for the photochemical PC board fabrication process.

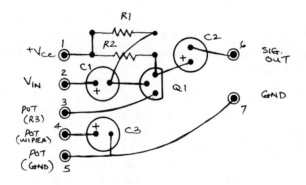

FIGURE 4-95 Rough sketch of example circuit

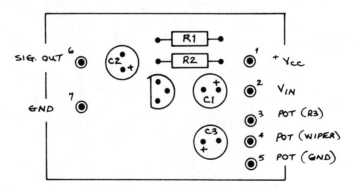

FIGURE 4-96 Component layout diagram for example circuit

216   Prototype Construction Projects

If the LPKF routing process is selected for fabricating the PC board, a different type of trace diagram is required. The rough sketch is still used to develop the rough sketch. Place tracing paper over the rough sketch which is taped on a light table *face up*. Because this machine can rout only in an X-Y (horizontal-vertical) plane, the trace diagram will display "islands" that represent the copper foil and horizontal and vertical lines that represent the insulation between the "islands." The 1:1 positive artwork required for the photochemical PC board process and the trace diagram for the LPKF routing system are compared in figure 4-97.

The PC board fabrication process follows the development of the 1:1 artwork master or the trace diagram. For the steps involved in the photochemical method, refer to Project 5 (Blinking LED Circuit).

If the LPKF routing system is used, consult the manufacturer's operation manual before attempting to use the machine.

EXERCISE 4: PCB AND FINAL CHASSIS ASSEMBLY

After the PC board is fabricated, the drill sizes for the various holes on the PC board are determined. A table of common PC board lead diameters, hole size ranges, and drill bit ranges is provided in this section. A table showing drill sizes commonly used for PC boards is also provided.

The PC board is loaded with the components after the holes are drilled. The component leads are preshaped so that they can be easily placed into the holes drilled in the board. The component layout diagram developed in Exercise 3 is used to determine the correct location of each component. After all of the components are mounted, the board is soldered.

Purpose

- To determine the proper drill size for each hole before performing the drilling operation.
- To prepare component leads before loading the components onto the PC board.
- To determine the proper sequence of steps to complete the chassis assembly.

Components & Equipment

- Drill index for numbered drills
- Micrometer
- Numbered drills (No. 68 to No. 50)

(A)

(B)

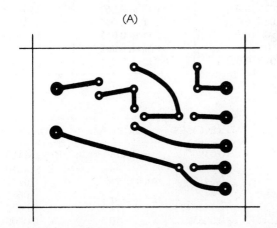
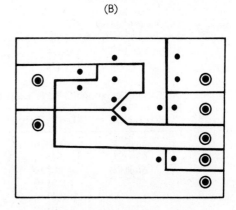

FIGURE 4-97 Comparison of positive artwork and LPKF system trace diagram (A) 1:1 positive artwork (B) LPKF trace diagram

Adjustable, Bipolar, Regulated Power Supply 217

- PC board drill press
- Miscellaneous soldering equipment (soldering station, solder stand, etc.)
- Hand tools (needle nose pliers, diagonals, nut drivers, screwdrivers, etc.)
- Chassis unit
- Circuit components

Procedure

The rough sketch developed in Exercise 3 will be used in this procedure to develop a *hole schedule diagram*. Locate the component with the smallest lead diameter to be mounted on the PC board. The *pilot drill* size will be determined by the lead diameter of this component; that is, *all* holes on the PC board will be drilled with this drill size.

To determine the pilot drill number, refer to Table 4-7. For example, if a transistor (TO-5 case) has the smallest diameter leads, the pilot drill number selected may be between 68 and 64.

Another method of determining the drill number to use for pilot holes or other component holes is to use a micrometer (an instrument used with a microscope or telescope to measure small distances) and Table 4-8.

NOTE: *The drill size selected by this method should be at least one size larger than the actual lead diameter (approximately 0.010").*

A much more convenient method of determining the drill size for any component lead is to use a *drill index* for the numbered drills.

TABLE 4-7 Numbered Drill Bit Range for Common Wire Sizes and Component Leads

WIRE SIZE OR COMPONENT	DIAMETER (IN INCHES)	HOLE SIZE (IN INCHES)	DRILL NUMBER
#18 Wire	0.0403	0.055 to 0.075	54 to 48
#20 Wire	0.0320	0.047 to 0.067	55 to 51
#22 Wire	0.0253	0.040 to 0.060	58 to 52
#24 Wire	0.0201	0.055 to 0.040	54 to 60
#26 Wire	0.0159	0.047 to 0.038	56 to 62
#28 Wire	0.0126	0.042 to 0.036	58 to 64
1/4 Watt Resistor	0.0250	0.040 to 0.060	58 to 52
1/2 Watt Resistor	0.0320	0.047 to 0.067	55 to 51
1 Watt Resistor	0.0400	0.055 to 0.075	54 to 48
2 Watt Resistor	0.0450	0.060 to 0.080	53 to 46
Disc Capacitors	0.0320	0.047 to 0.067	55 to 51
Miniature Electrolytics	0.025–0.032	0.047 to 0.067	55 to 51
TO-5 Transistors	0.016–0.021	0.031 to 0.036	68 to 64
DO-5 Diodes (1N914, Si)	0.0200	0.035 to 0.055	65 to 54
Rectifier Diodes (IN4000, Si)	0.0340	0.047 to 0.067	55 to 51
DIP, 14 & 16 Pin IC Sockets	0.017–0.021	0.032 to 0.036	67 to 64

TABLE 4-8 Common PC Board Drill Sizes

DRILL NO.	O.D. (INCHES)	DRILL NO.	O.D. (INCHES)	DRILL NO.	O.D. (INCHES)
68	0.0310	61	0.0390	55	0.0520
67	0.0320	60	0.0400	54	0.0550
66	0.0330	59	0.0410	53	0.0595
65	0.0350	58	0.0420	52	0.0635
64	0.0360	57	0.0430	51	0.0670
63	0.0370	56	0.0465	50	0.0700
62	0.0380				

Make a photocopy of your original rough sketch. This copy will be used to develop a *hole schedule diagram* which will be used when drilling the PC board. A hole schedule diagram is illustrated in figure 4-98 for the example circuit.

To drill the PC board:

1. Drill all of the holes with a number 60 drill.
2. Enlarge holes "A" with a number 33 drill.
3. Enlarge holes "B" with a number 52 drill.

After all of the holes are drilled on the PC board, install the turret terminals. Several types of tools and machines are available for mounting turret terminals and other terminals such as stakes, eyelets, and rivets. The tool most commonly used to fasten turret terminals is called a turret press (see figure 4-99 for the proper method of using this machine).

The procedure for the final chassis assembly is similar to those used in Projects 1 and 2. Review the basic process used in those projects before completing this phase of the project. A chassic assembly drawing for the power supply is provided to assist in the assembly process (see figure 4-100).

Use the power supply wiring diagram (figure 4-101) and the schematic diagram exclusively to complete the wiring exercise for your power supply. A written wiring schedule is not required for this final construction stage. A wiring schedule developed as in Project 1 prior to performing the actual wiring process may be helpful in planning the wiring activity for each section of the total circuit.

An example of a completed power supply is shown in figure 4-102.

EXERCISE 5: TESTING PROCEDURE AND SPEC SHEET DEVELOPMENT

The rewards received after all the effort and time spent on any project should be a project that not only looks attractive but also functions properly. It can be very frustrating when the unit tested does not function when power is first applied. The beginning student should realize that a malfunction is usually caused by minor oversight in construction rather than by a defective component. Most construction errors can be corrected quickly and simply. Once the unit is functioning as anticipated, it is necessary to determine how well it will perform. For this purpose, specific tests are performed and the results are documented.

This final exercise is included to help solve some of the common errors that may occur in the construction process. Also included are methods of testing to develop the specification sheet for this regulated power supply.

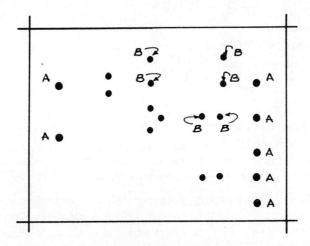

FIGURE 4-98 Hole schedule diagram for example circuit

FIGURE 4-99   Mounting turret terminals with turret press

Purpose

- To perform initial visual checks around the assembled unit before power is applied.
- To locate the probable cause of a circuit fault and to correct the malfunction.
- To develop and document the power specification by measurement with appropriate test equipment.
- To perform calibration adjustments to output voltage and meter circuitry.

Components & Equipment

- Completed power supply unit
- Hand tools
- Digital multimeter
- Oscilloscope
- 25-watt, variable load resistor (approximately 25 ohms)
- Miscellaneous test leads

Procedure

Initial Visual Check. The initial visual inspection is performed to determine if there are any missing wires or if any wires are terminated to wrong points. All hardware fasteners or other mounting hardware are inspected for proper fit and the components mounted on the chassis are checked for electrical isolation. The procedure for making this inspection follows.

1. Carefully check all interconnecting wiring between the mounted components on the front panel and chassis frame and the PC board or other destinations. This can be accomplished by comparing one wire at a time with the circuit wiring diagram or the schematic diagram. As each correct wire is verified, it should be checked off on the diagram.

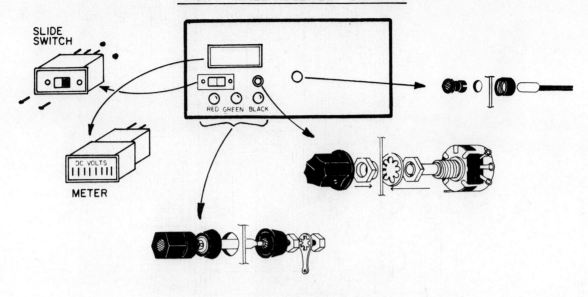
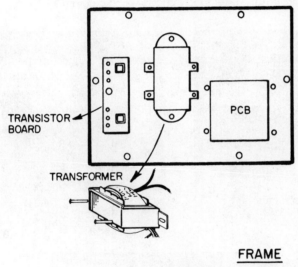
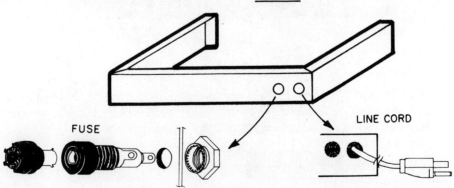

FIGURE 4-100 Power supply chassis assembly drawing

Adjustable, Bipolar, Regulated Power Supply

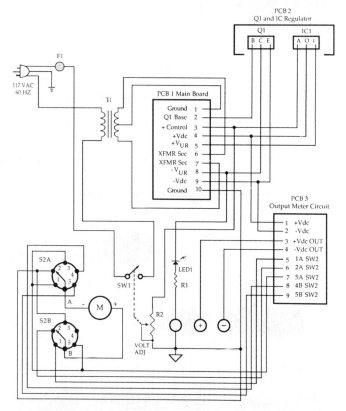

FIGURE 4-101 Power supply chassis wiring diagram

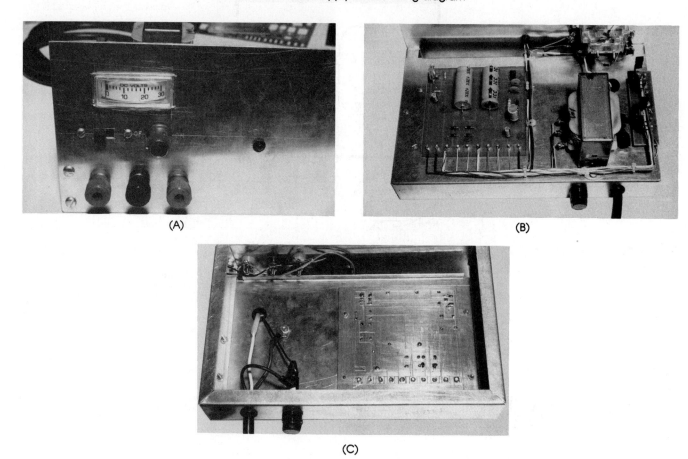

FIGURE 4-102 Completed power supply: (A) front view, (B) top view, (C) bottom view

2. Check the PC board for proper polarity of electrolytic capacitors and diode elements.
3. Check for any loose screws, nuts, or other fasteners or hardware that are too close or have touching electrical points.

Electrical Test on Chassis Mount Components. In this electrical test, the heat sink units on which the IC regulator and the power transistor (TIP125) are mounted are checked for possible connection to the chassis or other electrical points. Resistance readings across the output jacks to the chassis and across each other are also measured. The procedure for the electrical test follows.

1. A common problem encountered in the installation of the LM317 and TIP125 devices is an incorrect connection between the metal case of the package and the metal heat sink assembly to which it is mounted. An ohmmeter is used to check this connection. Connect one test lead to the chassis and bring the other lead in contact with the screw used to mount the device. If the device is properly installed, the meter will read *infinite resistance*. If any resistance indication or a short circuit (zero ohms) is measured, either the mica or fiber shoulder washer, or both, are not installed properly. Inspect the mica washer for any perforations that may have been caused by metal burrs around the rim of the mounting hole. Even a minor burr could cut through the washer. This problem can easily be solved by removing the burr and reassembling the device using a new mica washer and grease.
2. The output jacks are inspected next. Connect one lead of the ohmmeter to the chassis and bring the other lead in contact with each of the three output jacks separately. Again, if infinite resistance readings are recorded, the jacks were properly installed. If a short or specific value of resistance is recorded across any jack, check the insulating shoulder washer and replace it if it is defective or perforated.
3. The final ohmmeter check is to measure the resistance between the red (+) and black (−) jacks and also across each red and black jack to the green (ground) jack. Connect the negative (normally black) lead of the ohmmeter to the black jack. Make contact between the positive (normally red) lead and the red jack. This is the first resistance check. Refer to Worksheet 4-10 for the other combinations of ohmmeter test leads across the respective output jacks. Enter the resistance indication for each test on the worksheet.

WORKSHEET 4-10: Ohmmeter Test across Output Jacks

TEST POINTS	RESISTANCE INDICATION
1. Black ohmmeter lead to chassis; red test lead of ohmmeter to mounting screw of Q1(TIP125).	
2. Same hook-up as above. Red ohmmeter lead to mounting screw of IC1 (LM317).	
3. Black lead connected to chassis; red lead to +dc output jack.	
4. Same hook-up as no. 3; red lead to −dc output jack.	
5. Resistance between +dc jack and −dc jack: (A) Red lead to +dc jack; black lead to −dc jack. (B) Red lead to −dc jack; black lead to +dc jack.	

Adjustable, Bipolar, Regulated Power Supply

TEST POINTS	RESISTANCE INDICATION
6. Resistance between +dc jack and ground jack: (A) Red lead to +dc jack; black lead to ground jack. (B) Red lead to ground jack; black lead on +dc jack.	
7. Resistance between −dc jack and ground jack: (A) Red lead to −dc jack; black lead to ground jack. (B) Red lead to ground jack; black lead to −dc jack.	

Initial Power-on Test. With the power switch in the off position, plug the line cord into a 115-volt ac receptacle. Set the voltage-adjust pot (R2) fully counterclockwise to obtain the minimum output voltage. The meter selector switch should be on the voltage position. Turn the power switch on. The LED indicator lamp should be on and the meter should record approximately 2-volts dc. If these results are obtained, the power supply is functioning properly. If, however, these initial indications are not present, some troubleshooting tests must be performed. Several of the most common faults and their causes are listed in Table 4-9.

CAUTION: The power cord must be disconnected from the wall receptacle whenever repair work is being performed on the power supply.

TABLE 4-9 Troubleshooting Guide

FAULT	PROBABLE CAUSE
1. LED indicator lamp remains off.	(A) Check the polarity of the LED. (B) Defective LED. (C) LED series resistor value is too large.
2. Maximum output voltage with voltage adjustment at fully counterclockwise position; terminal voltage decreases as pot is turned clockwise.	(A) Wiring on R2 is reversed; switch outside connections on pot.
3. Fuse blows immediately when switch is turned on.	(A) Check the polarity of each diode on the PC board. (B) Check the polarity on the filter capacitors, C1 and C2. (C) Defective regulator, LM317. (D) Defective transistor, TIP125.
4. No output voltage adjustment.	(A) Check for missing or incorrect wiring on the adjustment terminal of LM317 and on R2 on the PC board.

Performance Test and Specification Sheet Development. The regulated power supply has a maximum output voltage range of 1.2-volts to 30-volts dc across the positive red jack and the negative black jack. The output is *bipolar* which means that equal values of two different polarities with respect to the green ground jack can be provided at the same time. This feature is attractive when working on op-amplifier circuits that require dual-supply voltages. If a bipolar supply is not available, two separate supplies connected in series must be used. If 30-volts dc is recorded across the positive and negative output jacks, the output voltages at the red and black jacks *with respect to the ground jack* is +15-volts dc and −15-volts dc. A single meter is provided on the front panel to monitor five functions, including three voltage functions and one current range.

Before the load current test is performed, the output voltage and meter calibration procedure, which follows, must be completed. The purpose of the output voltage balance adjustment is to insure that the

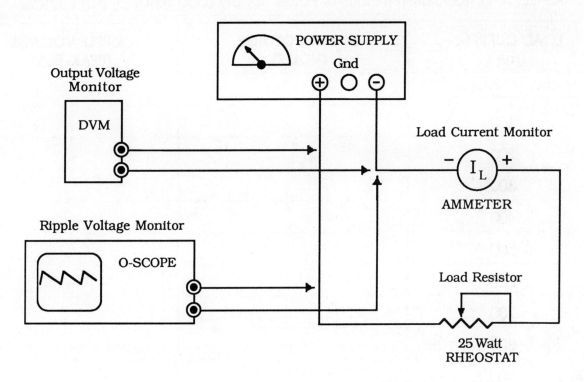

FIGURE 4-103 Power supply performance test circuit

positive and negative outputs are the same. This adjustment is very simple. It requires the use of the power supply meter on both the +V and −V functions.

Adjust the voltage to the full output across the red and black jacks with the panel meter in the proper position to monitor this voltage. Adjust this output voltage to *exactly* 30-volts dc. Rotate the meter switch to the +V position. Then, adjust the trim pot (R5) to indicate exactly +15-volts dc on the meter. Switch the meter to the −V position and observe the voltage indication. This negative voltage should follow the positive voltage.

If an imbalance between the positive and negative voltages is recorded, switch back to the initial meter range that recorded the 30-volts dc. A slight adjustment may be necessary to maintain this 30-volt setting. Repeat the +V and −V adjustments until the voltages are equal.

The ammeter ranges of the panel meter will be calibrated during the load current test. A second multimeter will be used exclusively to monitor load current.

The test circuit used to test the load current range of 0–1 amp, is shown in figure 4-103. An adjustable power rheostat capable of dissipating at least 50 watts is used as the circuit load. A digital meter is connected across the output jacks to monitor the regulation properties of this supply. An oscilloscope is also applied across the output at the same time to record the ripple voltage under various load conditions.

The purpose of the load current performance test is to see if the output voltage changes significantly and by what amount, with a change of load current over the range of no-load (0-amps) to full-load (1-amp). At the same time, the ripple voltage is measured and recorded to determine the change in ripple voltage for the same no-load to full-load conditions. These recorded values will later be used in equations that determine the two important power supply specifications—percent of regulation and percent of ripple.

Perform the load test and record the measured values for each load current value as indicated in Worksheet 4-11.

WORKSHEET 4-11: Tabulated Results of Power Supply Load Test

LOAD CURRENT (mA)	OUTPUT VOLTAGE (VOLTS)	RIPPLE VOLTAGE (PEAK-PEAK)
0		
100		
200		
300		
400		
500		
600		
700		
800		
900		
1000		

The important power supply specifications are the *load regulation percentage* and the *ripple percentage*. The load regulation percentage is a numerical interpretation of the change in output voltage between a no-load and a full-load condition. The ripple percentage is a numerical interpretation of the amount of ripple (ac) voltage that exists on the dc output voltage. The calculation of these two power specifications is explained in the following procedures.

To determine the regulation percentage, refer to Worksheet 4-11 which lists the output voltage changes as load current is increased from 0 mA to 1 amp. Following is the equation used to determine this specification

$$\% \text{ REGULATION} = \frac{V_{out}(\text{N.L.}) - V_{out}(\text{F.L.})}{V_{out}(\text{F.L.})} \times 100$$

If the no-load (N.L.) voltage is measured at 30-volts dc while at full-load (F.L.), the recorded voltage is 29.5-volts. The regulation percentage for this supply is computed to be

$$\% \text{ REGULATION} = \frac{30 \text{ V} - 29.5 \text{ V}}{29.5 \text{ V}} \times 100 = 0.017 = 1.7\%$$

The ripple percentage is also determined by the measured ripple voltage recorded in Worksheet 4-10. The equation for this specification is

$$\text{RIPPLE \%} = \frac{V_{rms}(\text{F.L.}) \text{ of Ripple}}{V_{out}(\text{F.L.})} \times 100$$

Where:

$$V_{rms}(F.L) \text{ of Ripple} = \frac{V_{p-p} \text{ of Ripple}}{2} \times 0.707$$

Where V_{pp} = measured peak-peak value of ripple voltage and V_{out} (F.L.) = measured dc output voltage.

The ripple voltage measured at 0 mA of load current (N.L.) is 0-volts p–p. At 1 amp (F.L.), the ripple voltage is recorded at 10-mV p–p. The full-load output dc voltage is 29.5-volts dc. Following is the calculation of the ripple percentage

$$\text{RIPPLE \%} = \frac{0.00354}{29.5 \text{ Vdc}} \times 100 = 0.00012 = 0.012\%$$

The power supply specification of percent regulation and ripple percent are important ratings to consider when sensitive electronic circuitry is connected to the power supply. Any load variation may cause transient voltage spikes to be generated and applied to devices not designed to withstand voltage surges. A well-designed power supply will act as a buffer against harmful voltage surges. Since the power supply is the common element between multiple load connections, its function is to provide the required voltage and current necessary under varying load demands without introducing noise (ripple) or harmful voltage or current surges to circuits.

Appendixes

A. SYMBOL REFERENCE GUIDE

The Symbol Reference Guide presented in this chapter is divided into fourteen categories:

1. Basic passive components
2. Electronic devices
3. Contacts, contactors, switches, and relays
4. Connection devices
5. Sources of electricity
6. Protection devices
7. Lamps and visual signaling devices
8. Acoustical devices
9. Rotating machinery
10. The transmission path
11. Ground symbols
12. Component combinations
13. Logic symbols
14. Miscellaneous symbols

The purpose of this guide is to allow easy identification of any symbol encountered in the schematic diagrams used in the Assignments and Projects in this text. For each type of symbol covered in the guide, variations of the symbol are presented, and the page on which more information can be found is referenced.

This guide does not cover all electronic symbols. However, with a basic knowledge of electronic symbology, it is possible to determine the meaning of any unknown symbol.

BASIC PASSIVE COMPONENTS   Page No.

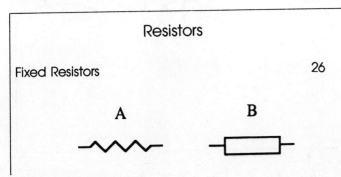

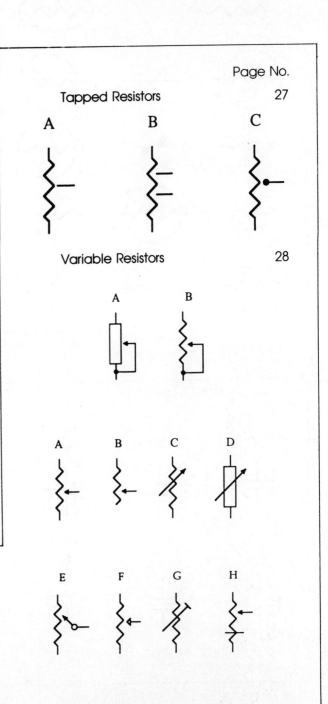

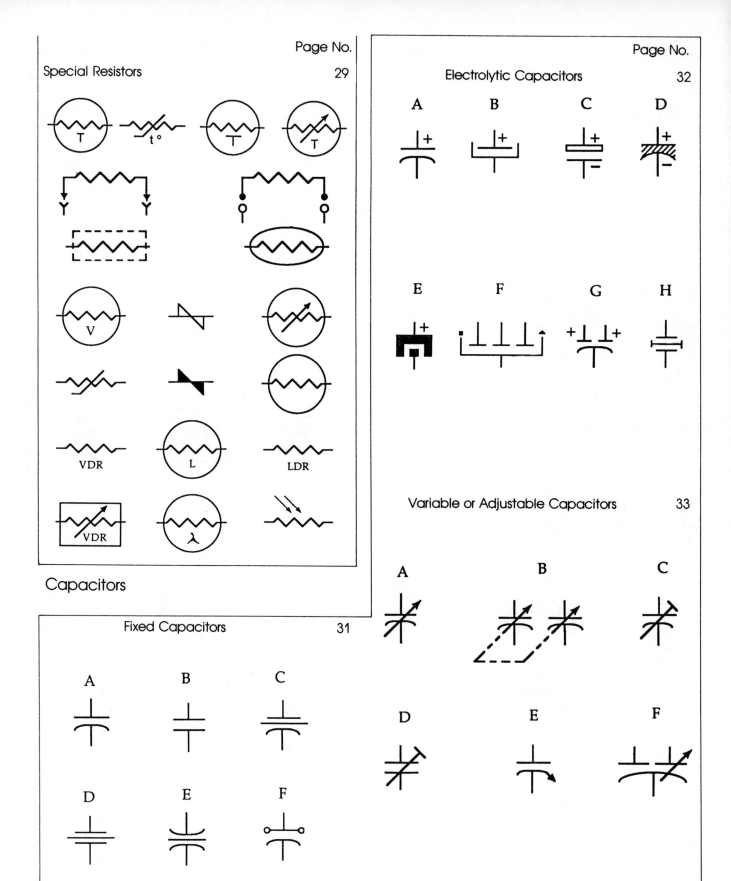

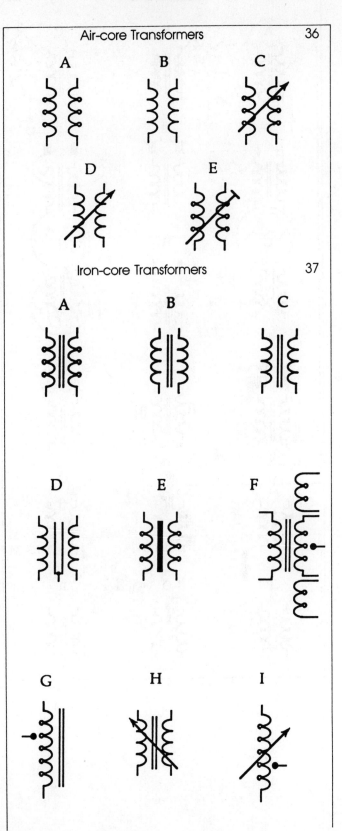

Powdered-iron-core Transformers	Page No. 37

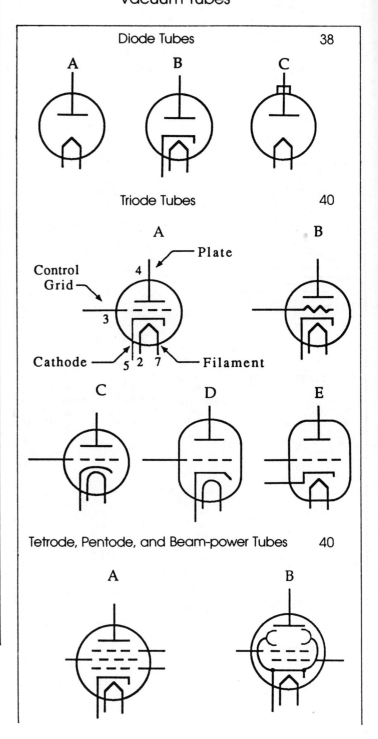

232 Appendix

Pentagrid Tubes   Page No. 41

Cathode-Ray Tubes   Page No. 43

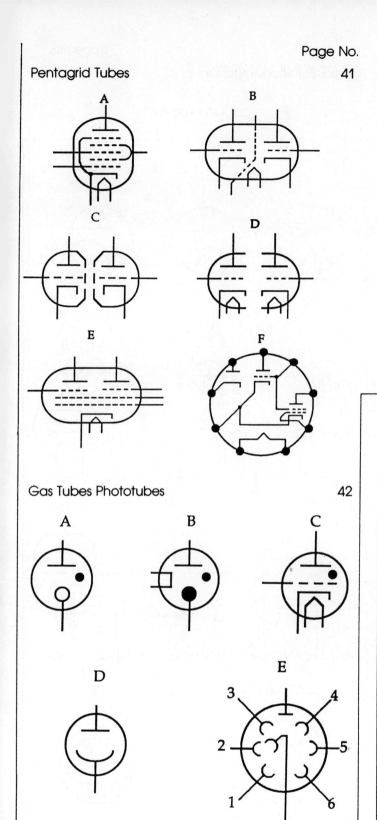
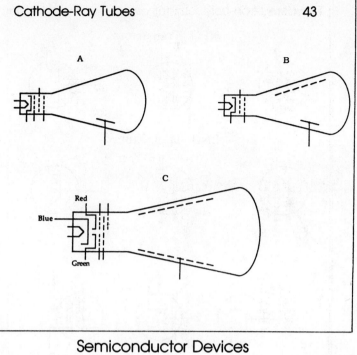

Gas Tubes Phototubes   42

Semiconductor Devices

Semiconductor Diodes   44

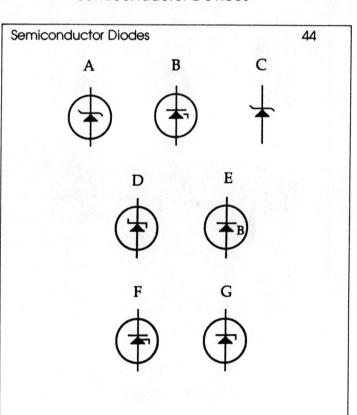

Appendix 233

Transistors Page No. 46

NPN Transistors

 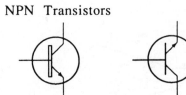

PNP Transistors

Special Semiconductor Devices Page No. 48

A B C

D E F

  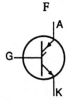

G H I

  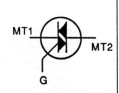

Page No. 47

A B C

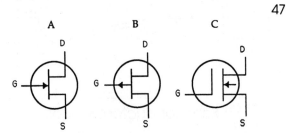

D E F

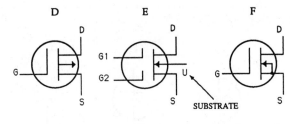

SUBSTRATE

TUNNEL DIODE

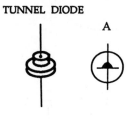

Page No. 49

A B C D

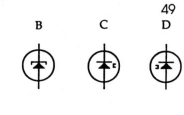

G H I

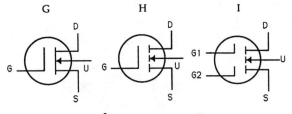

A B C D E F

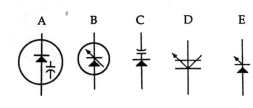

J K

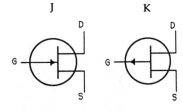

A B C

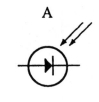 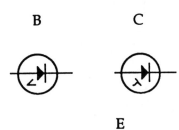

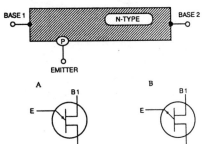

D E

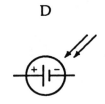

	Page No.
Integrated Circuits	50

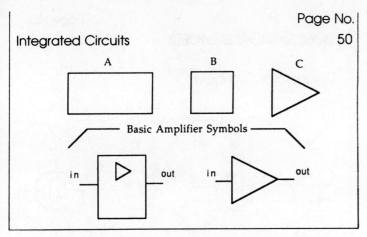

CONTACTS, CONTACTORS, SWITCHES, AND RELAYS

Switch Contact Identification

Single-Pole, Single-Throw (SPST) Contacts	52

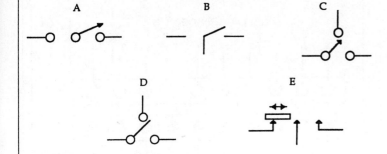

Single-Pole, Double-Throw (SPDT) Contacts	52

Double-Pole, Single-Throw (DPST) Contacts	53

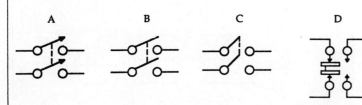

	Page No.
Double-Pole, Double-Throw (DPDT) Contacts	53

Wafer Switch	54

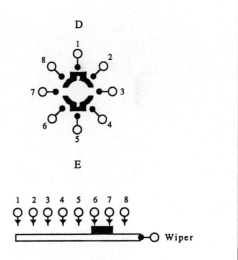

Appendix 235

	Page No.
Push-button Switches	55

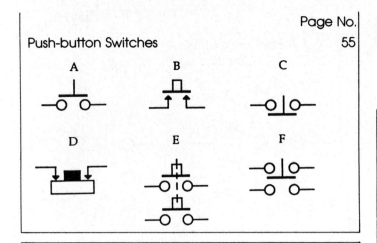

	Page No.
CONNECTION DEVICES	
Connector Symbols	

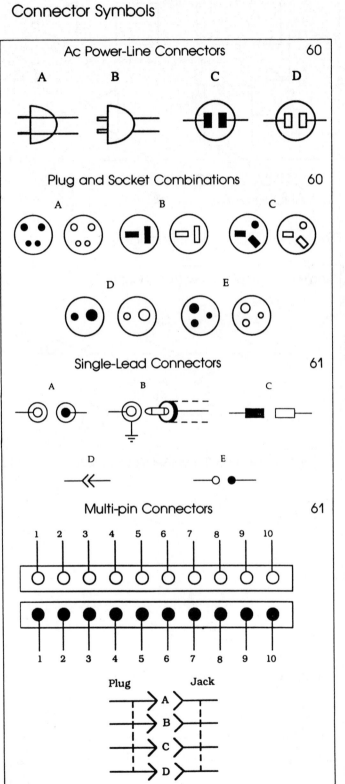

RELAYS 56

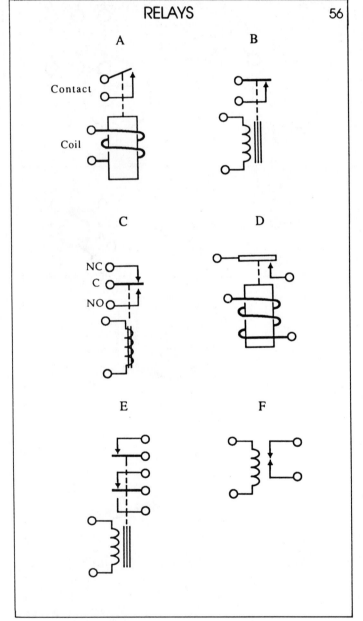

236   Appendix

Audio Connectors — Page No. 62

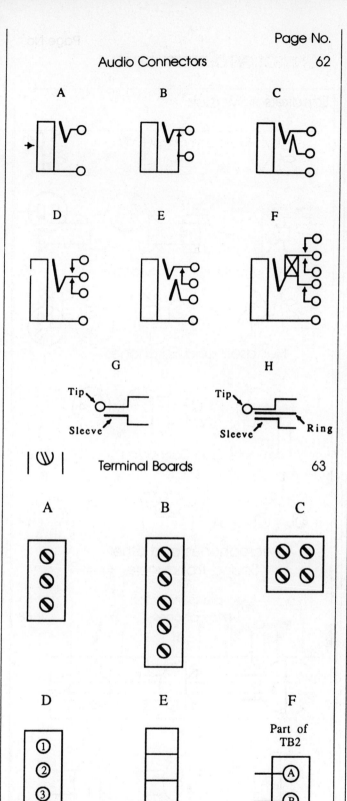

Terminal Boards — 63

SOURCES OF ELECTRICITY — Page No.

Direct Current (dc) Sources — 63

Alternating Current (ac) Sources — 64

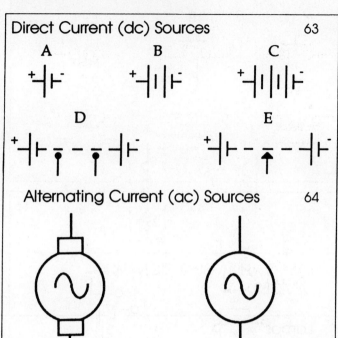

PROTECTION DEVICES

Fuses — 65

Circuit Breakers — 65

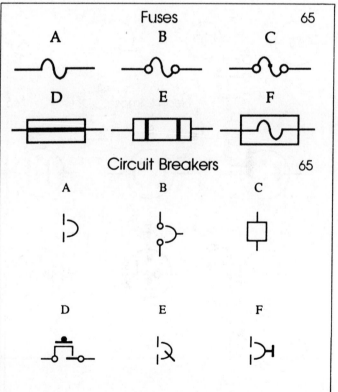

Appendix 237

Page No.

CIRCUIT BREAKERS (continued)

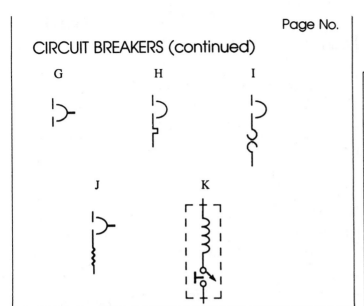

LAMPS AND VISUAL SIGNALING DEVICES

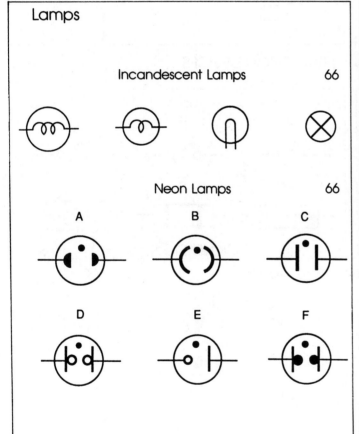

Page No.

ACOUSTICAL DEVICES

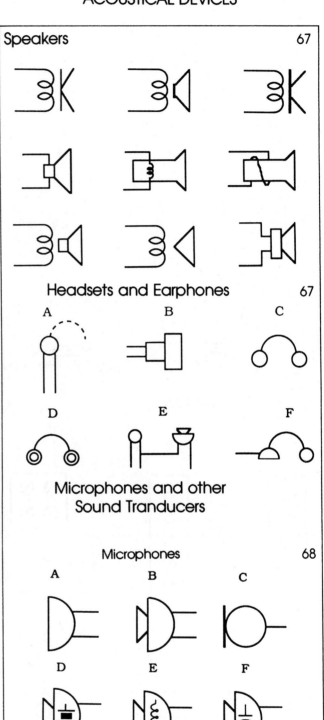

238   Appendix

Transducers

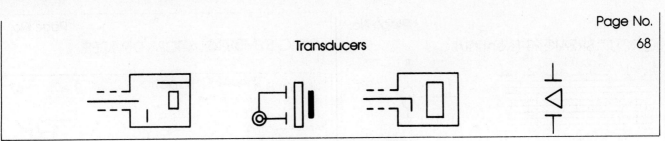

ROTATING MACHINERY

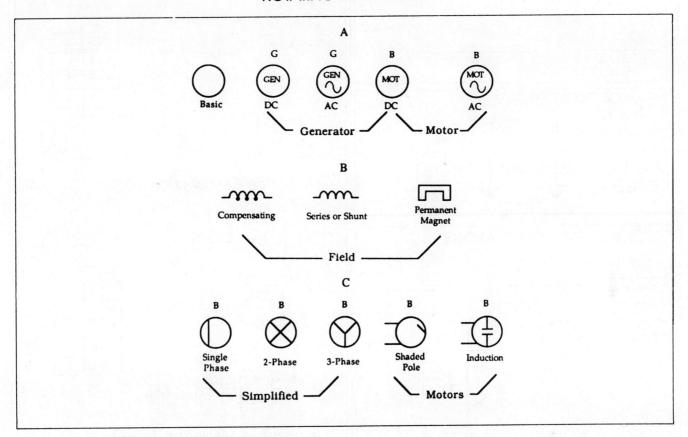

THE TRANSMISSION PATH

Connecting Wires and Cables

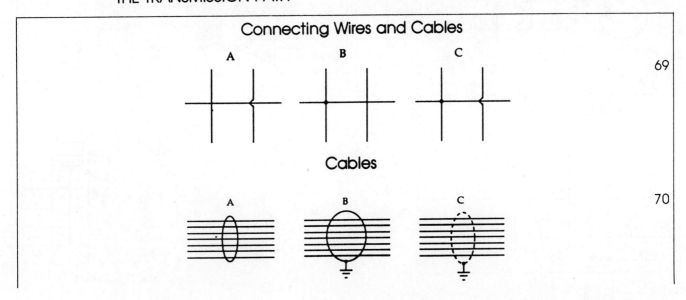

Cables

CABLES (continued)

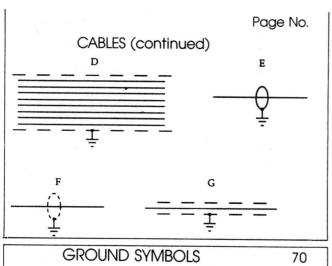

GROUND SYMBOLS 70

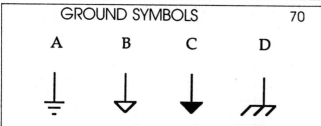

COMPONENT COMBINATIONS

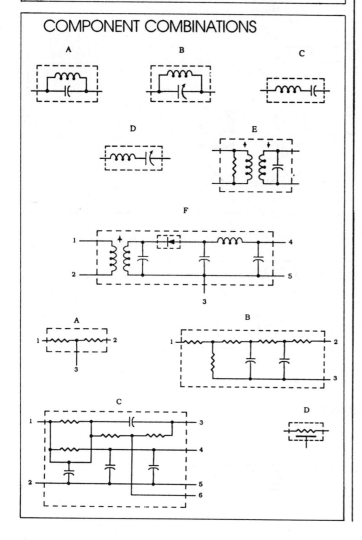

LOGIC SYMBOLS

The AND Function 74

The OR Function

The NOT and NAND Functions 74

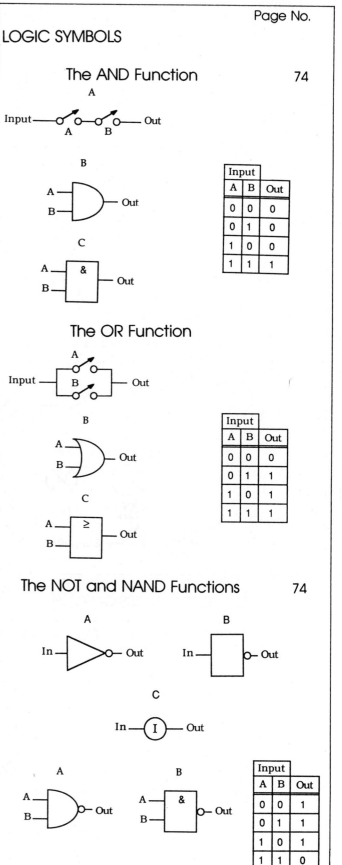

240 Appendix

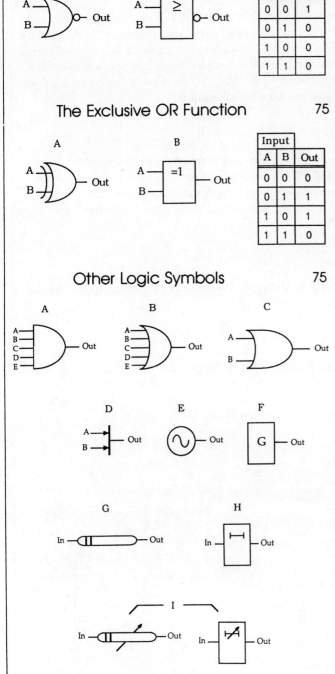
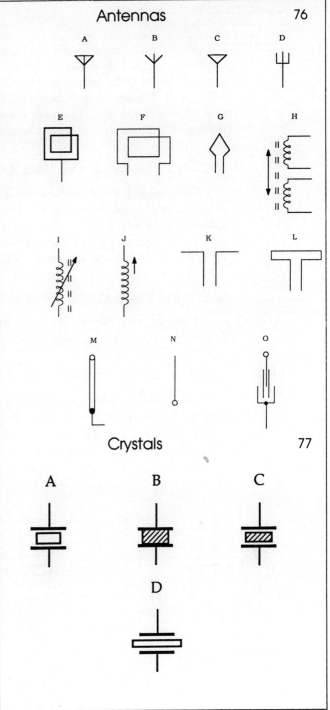

B. WIRE WRAPPING: THE TECHNOLOGY, METHODS, AND STANDARDS*

Introduction

As electronic industries have advanced in technology, the need for a faster, more reliable and inexpensive method of terminating wires to solderless terminal posts has become a necessity. In the past, it was sufficient to solder most connections since there was ample room between terminals and relatively few connections. The electronic circuitry used today, however, is more complex, smaller in size, and assembled in a reduced chassis area. This push towards high-density packaging has resulted in the technology of making connections to terminals by wrapping wire around terminal wrap posts.

The Technology of Wire Wrapping

Solderless wrapped connections are made by wrapping wire under tension a specified number of turns around a post with sharp corners. This method of connection was developed by Bell Telephone Laboratories, Western Electric Company. By bending the wire around the sharp corner of the terminal post, the oxide layer on both wire and terminal is crushed or sheared, and a clean, oxide-free, metal-to-metal contact is obtained (see figures A-1 and A-2).

Common Wrap Post Configurations

A terminal must have at least two sharp edges. The six types of wrap posts used most often today are illustrated in figure A-3.

Types of Wrapping Tools

There are three basic types of wire-wrapping tools used today: pneumatic tools, electric tools, and battery or hand operated tools. The pneumatic, or air, tool is preferred for production work. If an air compressor is not available, electric tools are recommended. Battery or hand operated tools are used predominantly for repair work or on smaller wiring projects (see figure A-4 through A-7).

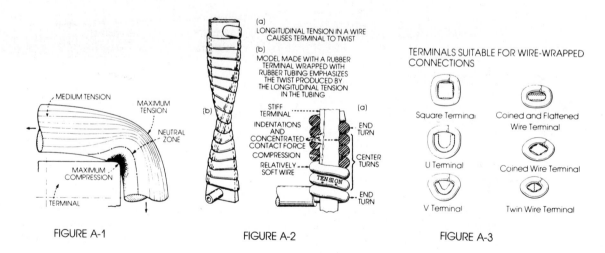

FIGURE A-1   FIGURE A-2   FIGURE A-3

*Reprinted, with changes, from "The Technology of Wire Wrapping," OK Industries Inc., catalog 84-36-T, 60-61, by permission of OK Industries Inc.

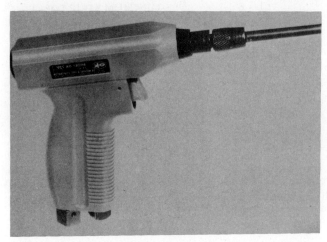

FIGURE A-4

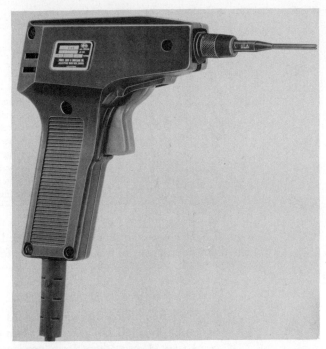

FIGURE A-5

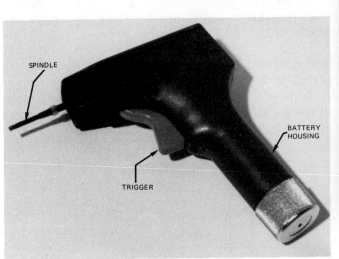

FIGURE A-6

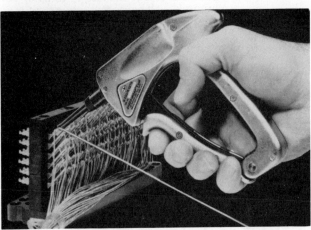

FIGURE A-7

Wire

The bare wire should be a solid, round, copper or copper-alloy conductor. It should be worked to the proper hardness to be indented by the wrap post. It must also be coated with a continuous and unbroken coating of tin, tin-lead alloy, silver, or gold. The quality of the conductor should be such that it is capable of being unwrapped without breaking or cracking.

The insulation must be removed manually or with an automatic wire stripper. Consult Table A-1 for the proper strip force when using automatic tools.

Appendix 243

TABLE A-1 Strip Force Chart

Awg	WIRE SIZE Diameter in Inches	Diameter in mm	MINIMUM NUMBER OF TURNS Class A (Modified Connection)	Class B (Conventional Connection)	MINIMUM STRIP FORCE lbs.	gms
16	0.051	1.30	4 stripped and ½ insulated	4 stripped	15	6800
18	0.0403	1.00	4 stripped and ½ insulated	4 stripped	15	6800
20	0.032	0.80	4 stripped and ½ insulated	4 stripped	8	3600
22	0.0253	0.65	5 stripped and ½ insulated	5 stripped	8	3600
24	0.0201	0.50	5 stripped and ½ insulated	5 stripped	7	3200
26	0.0159	0.40	6 stripped and ½ insulated	6 stripped	6	2700
28	0.0126	0.32	7 stripped and ½ insulated	7 stripped	5	2200
30	0.0100	0.25	7¼ stripped and ½ insulated	7 stripped	3.3	1500

The bare wire should not be exposed over a long period of time before the wrapping is performed. Only the amount of insulation required for the number of turns specified should be removed.

Types of Wrap

The strength of the wire-wrapped connection is considerably greater than that of a soldered connection. It is less easily stripped from the terminal and is less easily broken.

A *Class A*, or *modified* wrap, produces a wrap in which a portion of the insulation in addition to the bare wire surrounds the terminal. This type of wrap produces a mechanically and electrically stable connection. The number of turns of bare wire around the post depends on the gauge of wire used. A minimum of one-half turn of insulation around the wrap post is sufficient to insure better vibration characteristics. Severe vibrations are encountered by mobile equipment, equipment transported repeatedly, or by equipment subjected to other environments that exhibit appreciable vibrations, e.g., airborn applications.

A *Class B*, or *conventional* wrap, is the same as the modified wrap except that it has an additional half-turn of insulated wire around the wrap post. Class B wraps are used when the connection is in an environment where vibrations are negligible. Negligible vibrations include those encountered by stationary equipment at operational sites and during shipment of stationary equipment. Both types of wraps are illustrated in figure A-8.

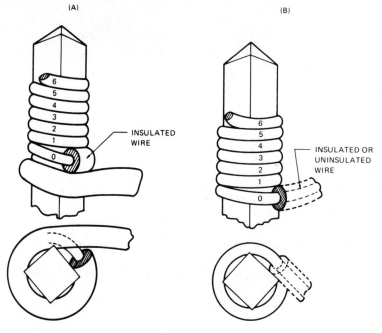

FIGURE A-8

244   Appendix

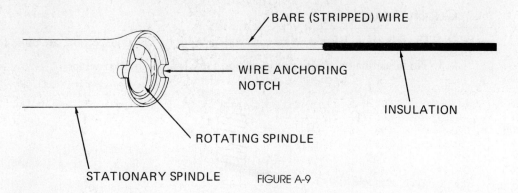

FIGURE A-9

Number of Wrap Turns

The correct number of turns to use when making Class A (modified) or Class B (conventional) connections is shown in Table A-1. Except for the first and last half-turns of bare wire, the maximum space between adjacent turns of bare wire should not exceed one-half the nominal diameter of the bare wire. This requirement only applies to the minimum number of turns specified in the table.

Performing Wire-wrapped Connections

1. Insert wire into feed slot in rotating spindle.

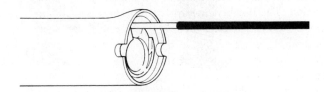

2. Anchor wire around wire anchoring notch.

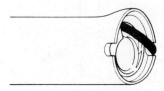

3. Insert into wrap post terminal.

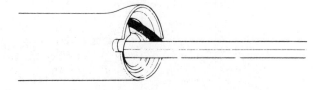

4. Finished connection. Class A modified wrap is illustrated.

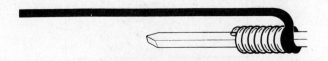

Quality Assurance

The following defects, which can be detected by a visual inspection, are cause for rejection:

1. No insulation or insufficient insulation for Class A modified wraps.

2. Improper wrapper spacing and overlapped wrappers. There should be no overlapping with the minimum specified number of turns of bare wire. An exception to this rule is when the same terminal wrap post is used for two or more wraps; that is, the first turn of insulated wire in a Class A modified wrap may overlap the last turn of bare wire in a connection below it on the same wrap post terminal.

3. Insufficient number of wrapper turns as indicated in Table A-1.

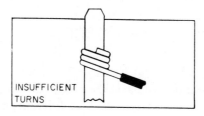

4. End tail does not conform to the preceding requirements.

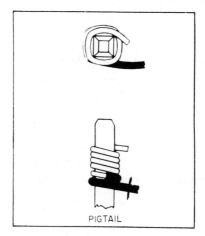

5. Overlapping wrapper turns within a wrapper level.

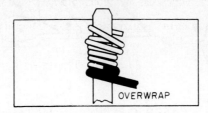

6. Space between adjacent wrapper turns exceeds one-half the diameter of the bare wire.

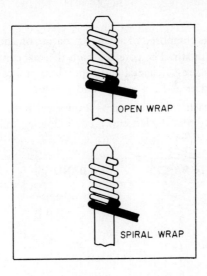

7. The wire should not be routed in a manner that will tend to unwrap any wire.

C. STANDARD RESISTOR COLOR CODE

A resistor is a component that will limit the current flow to a specific resistance when connected into an electrical circuit. Resistors have a total of 4 or 5 identifying bands. By looking at the colors, the resistance can be determined.

A *standard*, or composition, resistor is made of *graphite* and is coated with an insulating material (see figure A-10).

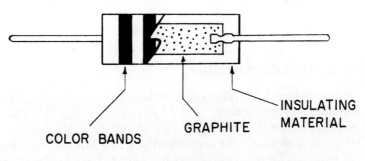

FIGURE A-10

Appendix 247

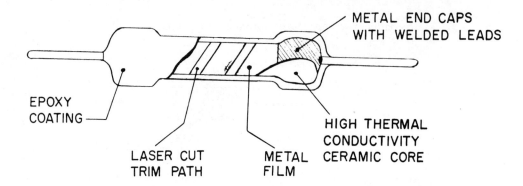

FIGURE A-11

A *metal film* resistor is composed of an epoxy coating on the outside and is wrapped with a metal film, which has a laser-cut trim path. The core is a high thermal-conductive ceramic material. The ends of the metal film resistor have a press-fit metal end cap. These resistors have a more precise tolerance than the composition resistors. Figure A-11 is a five-band resistor.

In the color code, each color represents a number. Resistor value is read from *left* to *right*. (Hold resistors so that colors begin at the left end.) The following table shows how the colors are used and the numbers they represent.

COLOR	1st BAND	2nd BAND	MULTIPLIER	TOLERANCE
BLACK	0	0	—	
BROWN	1	1	0	1%
RED	2	2	00	2%
ORANGE	3	3	000	
YELLOW	4	4	0,000	
GREEN	5	5	00,000	.5%
BLUE	6	6	000,000	.25%
VIOLET	7	7	0,000,000	.10%
GRAY	8	8	00,000,000	.05%
WHITE	9	9	000,000,000	
GOLD			.1	5%
SILVER			.01	10%
NO COLOR				20%

*Multiplier—the number of zeros to add.
*Tolerance—the percentage varience in a resistance.

NOTE: RESISTORS NOW COLOR-CODED ARE DONE SO ON THE BASIS OF ONE OF THREE METHODS.
  1. *Four-band configuration.*
  2. *Five-band configuration. (The only difference between these two methods is the "terminal coding" band on Method 2. The tolerance coding is the same for both methods.)*
  3. *Five-band configuration. (This method deviates from the preceding methods in allowing for a third significant figure, and in using a different tolerance coding system.)*

Reading Four-band Resistors

1. One near left end, first digit (direct value).
2. Second digit (direct value).
3. Number of zeros to add after the first two figures (multiplier).
4. (If present), on right is the tolerance band which indicates the tolerance range for the total value of the resistor. This band will always be gold (± 5%) or silver (± 10%). If a tolerance band is not

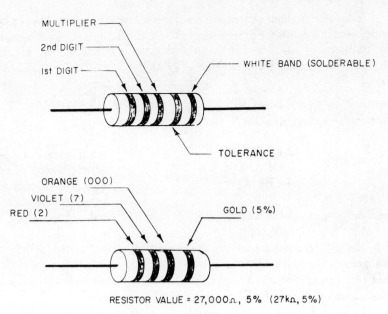

FIGURE A-12

present, the tolerance is 20%. There may be a fifth white band on the far right. This band is *not* a tolerance band. This is a vendor coding, meaning that the leads are solderable. It is read as a four-band resistor, as the white band is not used.

Examples of four-band resistors are shown in figure A-12.

Reading Fifth-band Resistors

1. One near left end, first digit (direct value).
2. Second digit (direct value).
3. Third digit (direct value).
4. Number of zeros to add after the first three figures (multiplier).
5. Tolerance band and can be any of the tolerances indicated on the color code chart.

Examples of five-band resistors are shown in figure A-13.

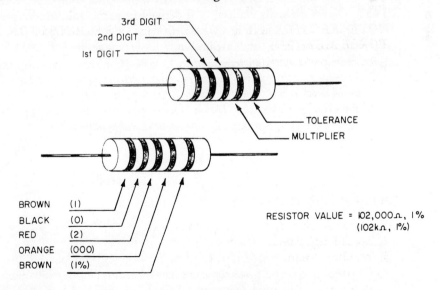

FIGURE A-13

Appendix 249

The ability to limit current in a resistor is measured in *ohms*. The symbol is the Greek letter omega (Ω) and is used in a resistance of less than 1,000. Contractions are used instead of actually writing all the zeros for resistance of 1,000 to 1,000,000 (the measure in kilohms or "k"), and for resistance of 1,000,000 or more (the measure being megohms or "M").

units	0-BLACK	EXAMPLE:	
hundreds	1-BROWN		
thousands "k"	2-RED	red, red, green, silver	2.2M ± 10%
	3-ORANGE	white, green, orange, red, green	95.3k ± .5%
	4-YELLOW	yellow, violet, green	4.7M ± 20%
		white, black, red, gold, white	9k ± 5%
		brown, gray, black, silver	18Ω ± 10%
millions "M"	5-GREEN	red, yellow, green, black, brown	245Ω ± 1%
	6-BLUE		
billions "G"	7-VIOLET		
	8-GRAY		
	9-WHITE		

The physical size of resistors is directly proportional to the amount of wattage (heat) they are built to withstand. The four most common sizes are shown in figure A-14. On four-band, composition resistors, you can determine the part number by knowing the wattage and the value of the color coded bands.

WATTAGE	4th BAND
¼	GOLD
¼	SILVER
½	GOLD
½	SILVER
1	GOLD
1	SILVER
2	GOLD
2	SILVER

The wattage and tolerance band of a resistor determines the prefix number. The color coded bands on the resistor determine the suffix number. For instance, a ¼ watt resistor with a gold fourth band is prefixed 0683. The suffix number, for a ¼ watt resistor whose color coded bands are brown, yellow, orange, gold is 1435. The fourth suffix number, 5, is derived from the gold band representing ± 5%.

A ¼ watt resistor, with a silver fourth band, is prefixed 0684. The suffix number, for a ¼ watt resistor whose color code bands are brown, yellow, orange, silver is 1431. The fourth suffix number, 1, is derived from the silver band representing ± 10%. (Use the 1 and drop the 0.)

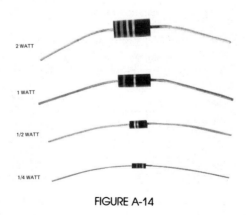

FIGURE A-14

D. INDUSTRIAL RECTIFIER OUTLINES*

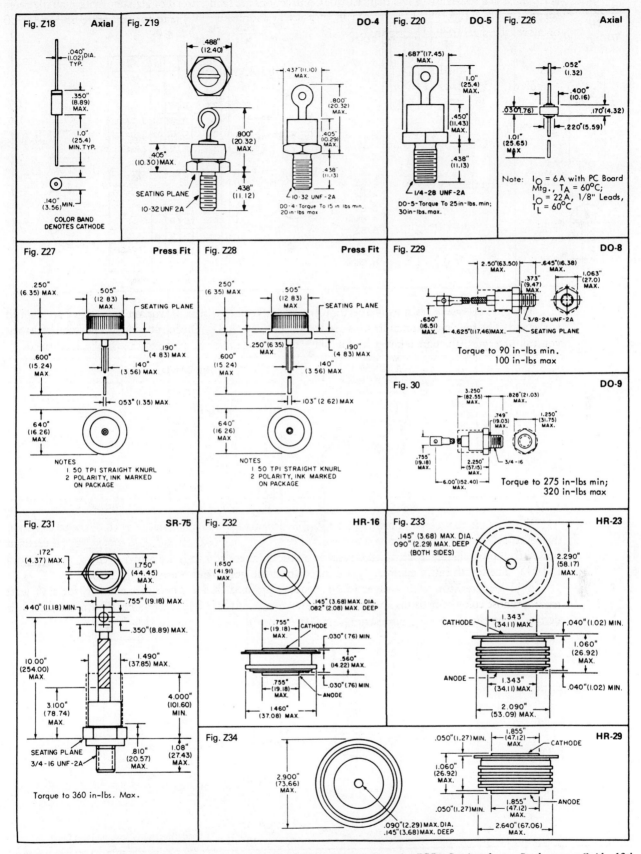

*The figures on this page and pages 237–247 are reprinted with permission from the *ECG® Semiconductor Replacement Guide*, 12th Edition. ECG® is the registered trademark of Philips ECG, Inc., a North American Philips Company.

E. TRANSISTOR OUTLINES

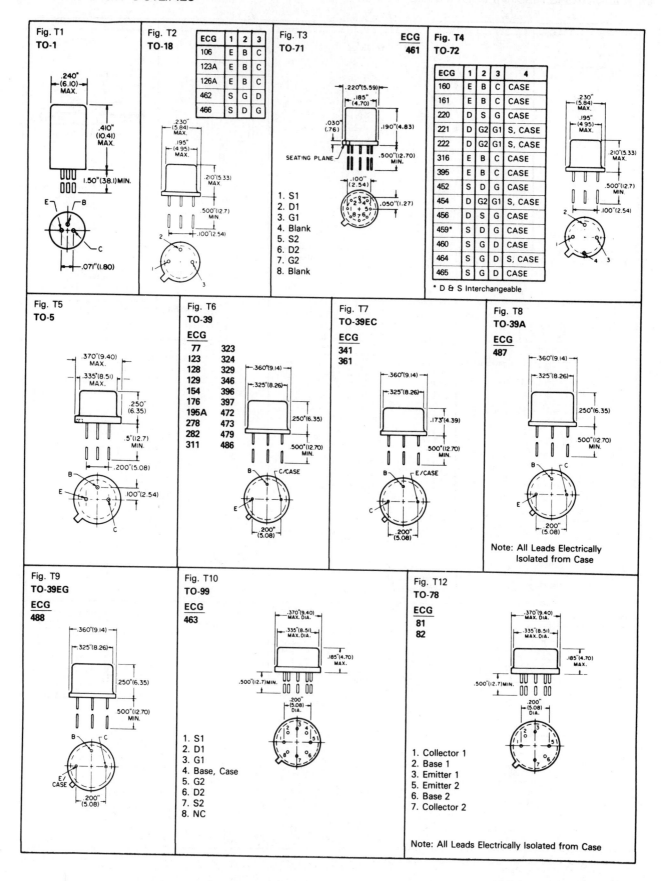

252 Appendix

TRANSISTOR OUTLINES (continued)

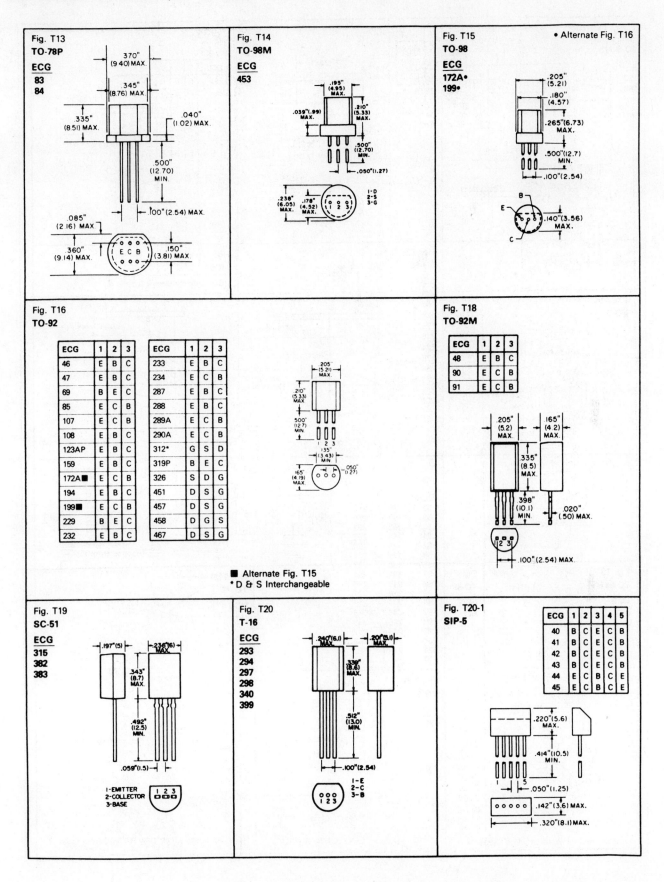

Appendix 253

TRANSISTOR OUTLINES (continued)

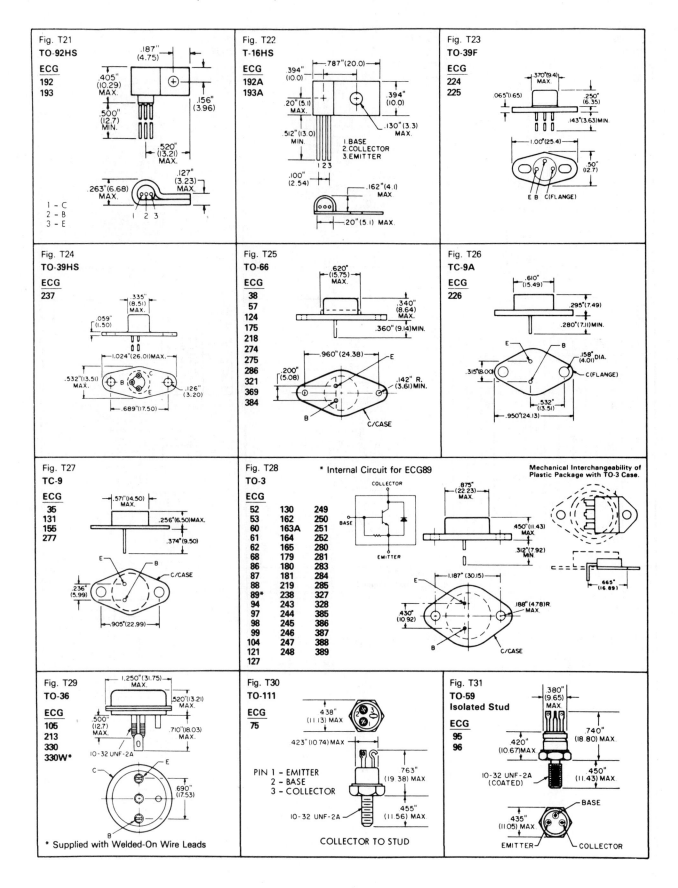

254 Appendix

TRANSISTOR OUTLINES (continued)

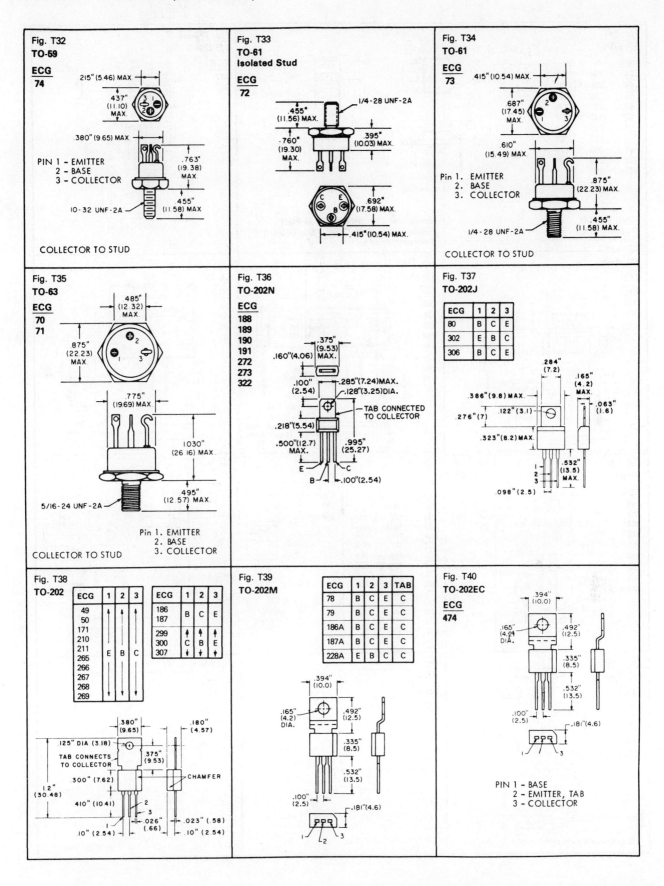

Appendix 255

TRANSISTOR OUTLINES (continued)

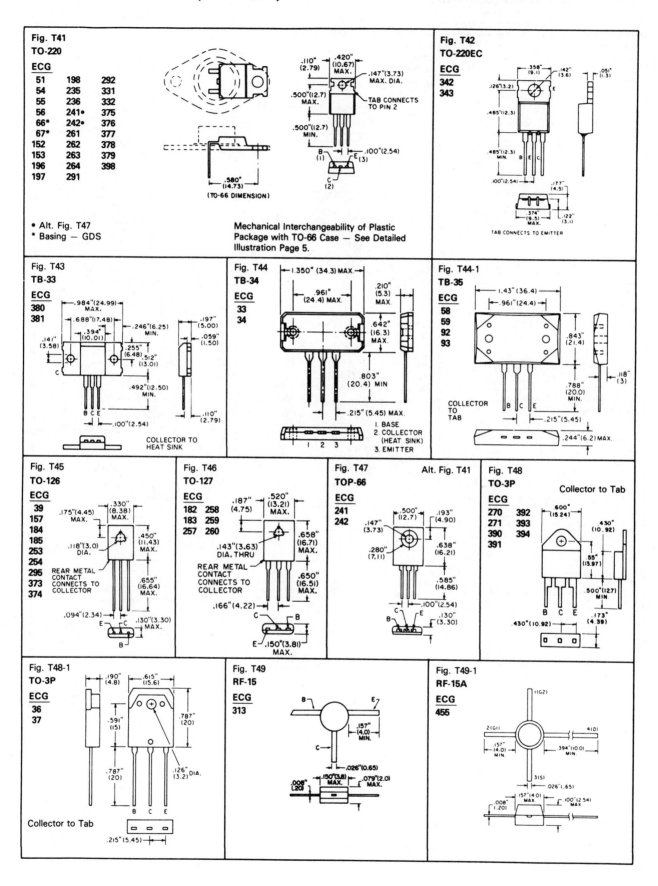

TRANSISTOR OUTLINES (continued)

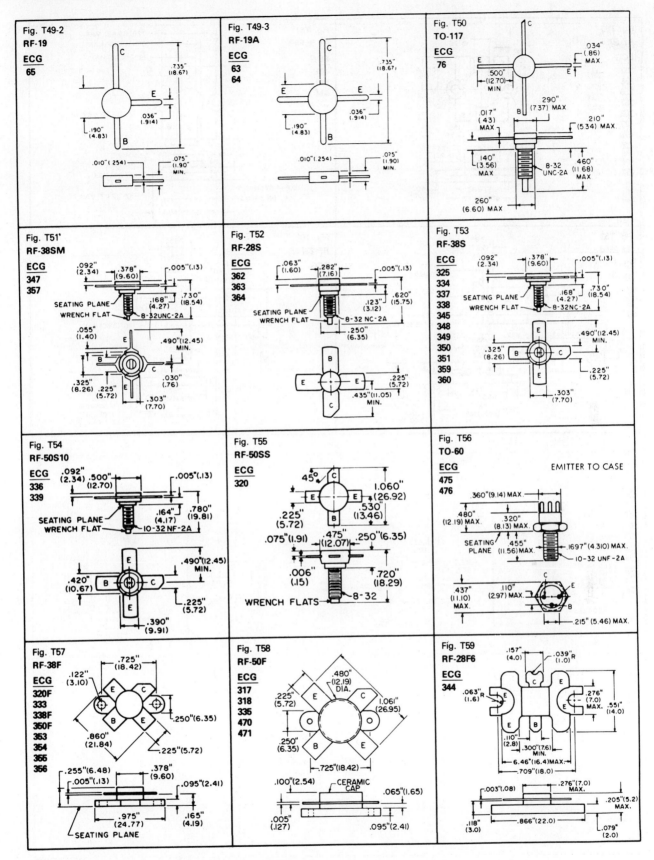

Appendix 257

TRANSISTOR OUTLINES (continued)

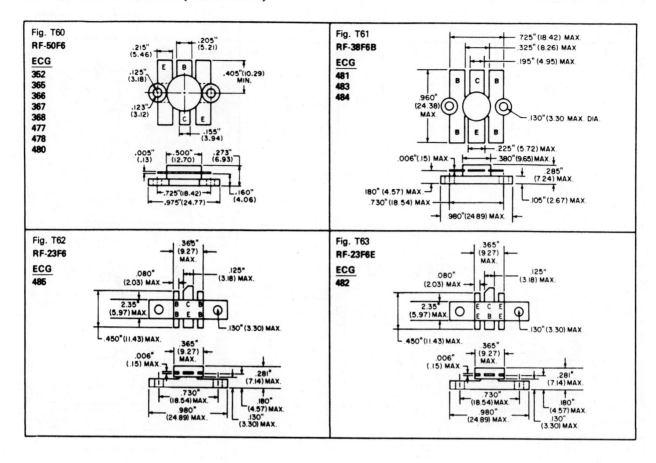

F. SCR OUTLINES

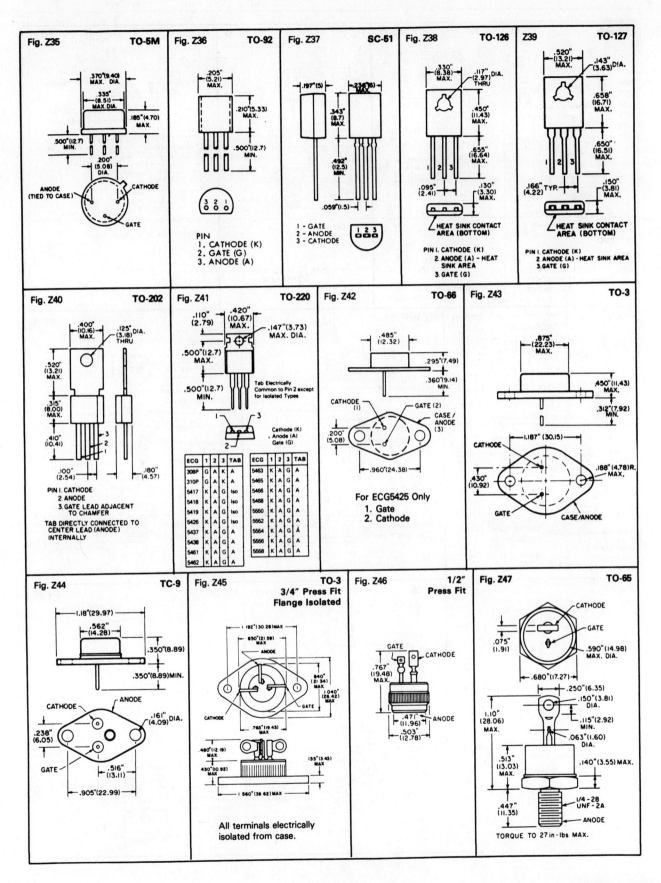

Appendix 259

SCR OUTLINES (continued)

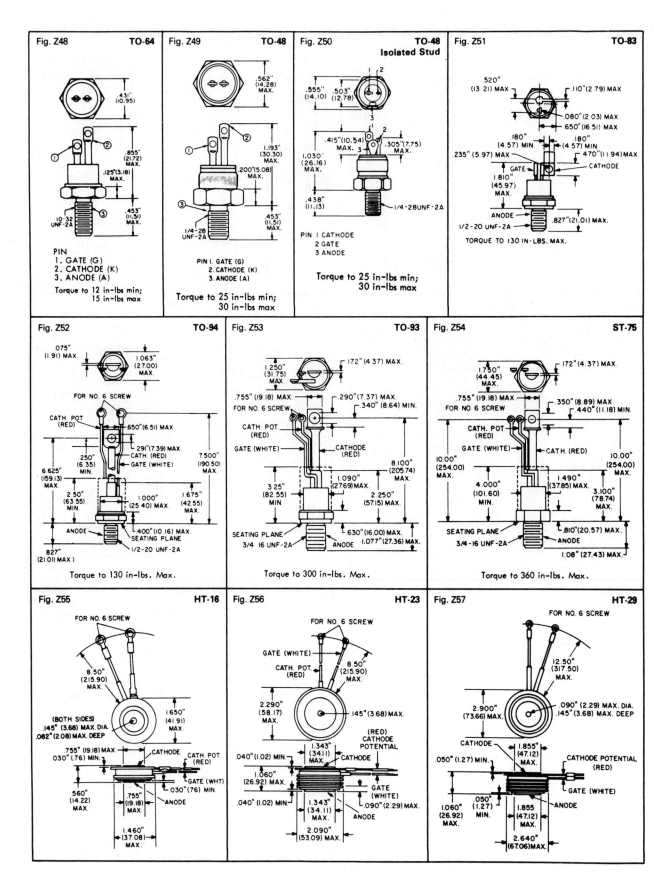

260 Appendix

G. TRIAC OUTLINES

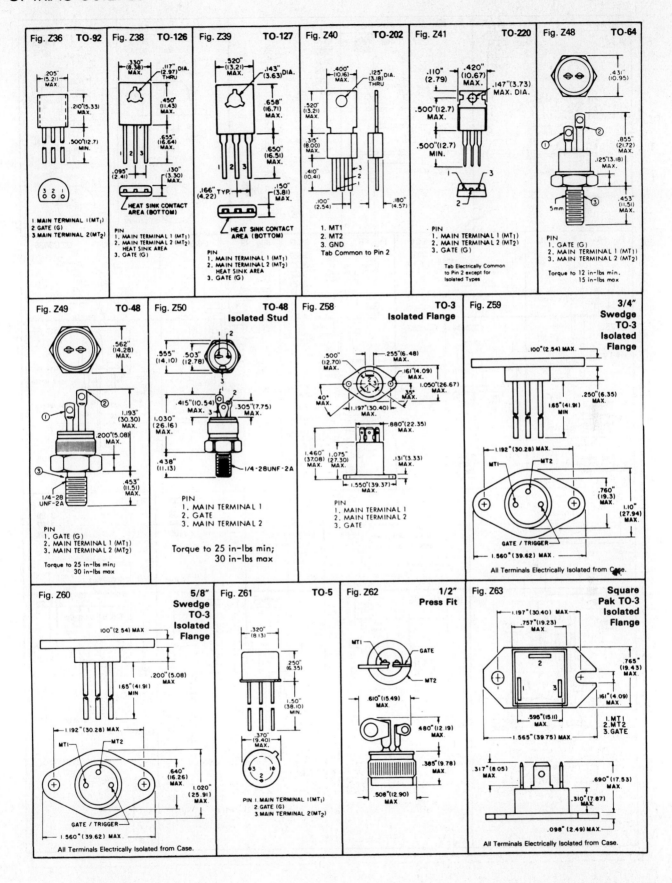

Appendix 261

H. DISCRETE LED INDICATOR OUTLINES

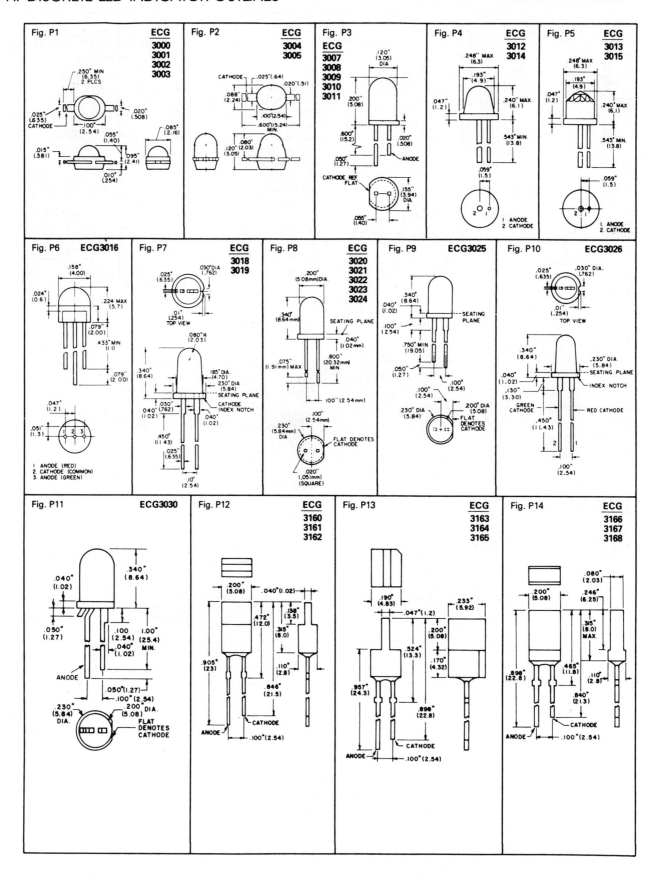

262 Appendix

I. INTEGRATED CIRCUIT STANDARD PACKAGE

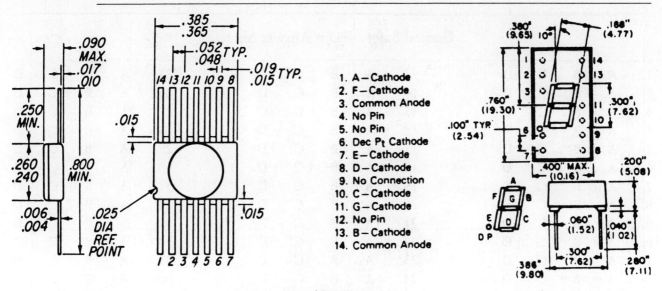

1. A – Cathode
2. F – Cathode
3. Common Anode
4. No Pin
5. No Pin
6. Dec P$_t$ Cathode
7. E – Cathode
8. D – Cathode
9. No Connection
10. C – Cathode
11. G – Cathode
12. No Pin
13. B – Cathode
14. Common Anode

MAN 7
Common Anode LED Display

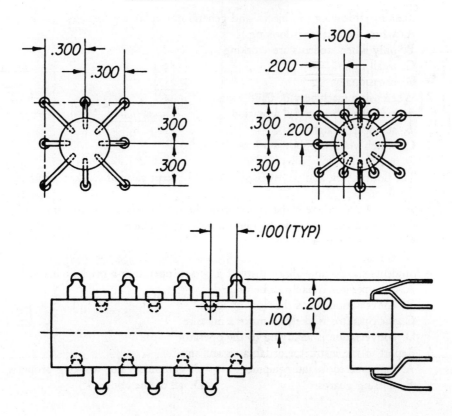

Appendix 263

NAME

General Safety Exam Answer Sheet

1.	A	B	C	D	13.	A	B	C	D	25.	A	B	C	D
2.	A	B	C	D	14.	A	B	C	D	26.	A	B	C	D
3.	A	B	C	D	15.	A	B	C	D	27.	A	B	C	D
4.	A	B	C	D	16.	A	B	C	D	28.	A	B	C	D
5.	A	B	C	D	17.	A	B	C	D	29.	A	B	C	D
6.	A	B	C	D	18.	A	B	C	D	30.	A	B	C	D
7.	A	B	C	D	19.	A	B	C	D	31.	A	B	C	D
8.	A	B	C	D	20.	A	B	C	D	32.	A	B	C	D
9.	A	B	C	D	21.	A	B	C	D	33.	A	B	C	D
10.	A	B	C	D	22.	A	B	C	D	34.	A	B	C	D
11.	A	B	C	D	23.	A	B	C	D	35.	A	B	C	D
12.	A	B	C	D	24.	A	B	C	D	36.	A	B	C	D

The laboratory safety procedures have been read and explained to me. I have also taken and have satisfactorily passed the written safety exam. I agree to abide by these rules, and when in doubt, I will ask the instructor.

_____   _____
Student's Signature   Date

J. ELECTRICAL SAFETY EXAM

Read each question carefully. Indicate your responses on the answer sheet provided. There is only *one* correct answer for each question.

1. Running, throwing of objects, and general horseplay are forbidden activities in the shop:
    A. when the teacher is looking
    B. only when students are working
    C. at all times
    D. occasionally
2. When using machinery or hand tools:
    A. give the job your full attention
    B. stand up straight
    C. watch your neighbor working
    D. watch the time go by
3. The floor, aisles, and passageways should be kept clear of stock, tools, and materials. Objects on the floor
    A. may be left there if the operator of the machine is in a hurry to leave
    B. may cause someone to trip into a moving machine
    C. may be ignored
    D. are unsightly
4. Students should not talk or distract a person operating a machine because
    A. the operator is likely to be injured
    B. conversation slows the flow of work
    C. the operator is likely to make a mistake
    D. conversation is annoying to the operator
5. Report to the instructor or lab assistant any
    A. damaged tools and equipment
    B. missing guards
    C. equipment not working properly
    D. all of the above

6. Never operate lab equipment when a certified instructor is
   A. helping another student
   B. in another area of the lab
   C. not scheduled (assigned) to be present
   D. none of the above
7. Most tools are designed for a specific use or purpose. If they are used incorrectly, the result may cause:
   A. damage to the student's project
   B. breakage of tools
   C. injury to the self or other students
   D. all of the above
8. Long hair is dangerous around rotating machinery. If it is long enough to get caught in the machinery, it must be:
   A. tied up and back
   B. trimmed off
   C. pulled out
   D. none of the above
9. Loose clothing must be securely fastened or removed and long sleeves rolled up above the elbows:
   A. before operating any machinery
   B. after operating the machinery
   C. during the operation of the machinery
   D. only when you are assisting the instructor
10. All accidents and injuries, no matter how slight must be
    A. ignored
    B. reported to the instructor or lab assistant immediately
    C. reported to the school nurse
    D. reported to another student
11. Only the operator and _____ are permitted within the working area around the machine.
    A. one other student
    B. the instructor
    C. a helper
    D. all of the above
12. Gasoline, paints, solvents, and other flammable materials should be used
    A. with another student
    B. in a well-ventilated area
    C. at the workbench
    D. in an enclosed area
13. Students are to operate only machinery or equipment for which they are/have _____ to operate.
    A. qualified
    B. permission
    C. instructions
    D. all of the above
14. When touching electrical switches, plugs, or receptacles, be sure your hands are dry because
    A. a switch will not operate if your hands are wet
    B. a plug will easily slip from your fingers if your hands are wet
    C. if your hands are wet, you may receive a severe electrical shock and/or serious burns
    D. none of the above
15. Acid or chemicals on your hands or face should be immediately washed away with plenty of
    A. water
    B. glycerine
    C. olive oil
    D. vaseline
16. If you notice any breakage or damage to tools, instruments, or machinery, you should
    A. report the damage immediately
    B. be careful when using such equipment
    C. say nothing because you might get the blame
    D. none of the above
17. If you are in doubt about the use of any tool, machine, equipment, or lab procedures
    A. ask an advanced student for help
    B. proceed cautiously
    C. always ask your instructor or a qualified aide
    D. none of the above
18. Eye protection is used to
    A. improve vision
    B. prevent flying particles or corrosive substances from entering your eyes
    C. prevent eyestrain
    D. none of the above

19. When tools are carried in the hands, keep the cutting edge or sharp points
    A. directed toward the floor
    B. directed away from the body
    C. directed over the head
    D. directed toward the body to protect others
20. Extension and power cords should always be checked and kept in good condition because
    A. breaks and tears in the cord are unsightly
    B. breaks and tears in the cord can cause serious shocks or burns
    C. sparks may cause wood to burn
    D. a short may cause the machines to burn up
21. Carbon dioxide (CO_2) fire extinguishers may be used to put out what type of fires?
    A. Electrical fires only
    B. Electrical and oil fires
    C. Oil fires only
    D. Any type of fires
22. Water should *never* be used to put out what type of fire?
    A. Wood fires
    B. Electrical and oil fires
    C. Paper fires
    D. None of the above
23. The proper procedure for fighting a fire with a fire extinguisher is to
    A. point the nozzle at the top of the flame
    B. point the nozzle at the middle of the flame
    C. cover the area around the fire and keep it from spreading
    D. point the nozzle at the source of the fire because that is where the fire is located
24. Lifting any object too heavy for you
    A. is all right if you do it slowly
    B. can be done if you know the right way to lift
    C. should never be done because it may cause strain or rupture
    D. is a good way to show off your strength
25. Deliberately shorting an electrical circuit
    A. is permissible if the voltage is low
    B. may damage the wires
    C. is an easy method of testing whether the circuit is closed or open
    D. may cause an explosion or do bodily harm
26. Cutting two or more "hot" wires with pliers
    A. is a safe practice if the handles of the pliers are insulated
    B. is permissible if the wires are 18-gauge
    C. may be done safely if you are standing on a wooden floor
    D. none of the above
27. Laboratory area and equipment clean-up is the responsibility of
    A. the custodian
    B. all students
    C. the instructor
    D. the parts person
28. Be sure your hands are free as much as possible of _____ before using hand tools.
    A. dirt
    B. grease
    C. oil
    D. all of the above
29. Major repairs on any lab machinery or equipment may be performed with:
    A. the power on
    B. the instructor's permission
    C. the machine running
    D. all of the above
30. Spilled oil, grease, or unknown liquids are dangerous. Always
    A. clean it up after identifying it
    B. pour water on it
    C. leave it
    D. none of the above

31. The motion involved in striking or cutting must be done in a direction
    A. towards you
    B. towards other students
    C. away from you
    D. all of the above
32. A project is still dangerous even after its power switch is turned off because
    A. it may still be plugged in
    B. some of the components may be "hot"
    C. the capacitors can store a charge which can cause an uncomfortable shock
    D. all of the above
33. Never use a file
    A. without a handle
    B. by throwing it
    C. as a pry bar
    D. all of the above
34. Pass tools to another student
    A. with handles first
    B. with the point first
    C. by throwing them
    D. none of the above
35. Before starting a machine, you must
    A. check all adjustments
    B. make sure all guards work
    C. remove all tools/rags
    D. all of the above
36. Before leaving a machine, you must make sure
    A. that the guards are off
    B. that the machine has come to a complete stop
    C. that the power is off
    D. both (B) and (C)

Index

Acoustical devices
    headsets and earphones, 67
    microphones, 67
    speakers, 66
    transducers, 68
Alternating current (ac)
    sources, 64
Antennas, 76
Assembly
    cable (Assignment 4), 100
    chassis (Project 2), 140
    printed-circuit board (Assignment 5), 108
    terminal strip (Assignment 3), 92

Batteries and cells, 63
Bipolar-junction transistors (BJT)
    active elements, 45
    NPN or PNP construction, 45
BNC connector. See Coaxial cable assembly
Breadboarding technique, 189

Cable assembly
    coaxial, 103
    inspection and testing, 106
    TSP (Twisted-Shielded-Pair), 100
Capacitors, types of
    combinations, 72
    electrolytic, 31
    fixed, 30
    variable, 33
Chassis
    assembly (Project 2), 140
    fabrication (for Project 2), 130
    planning and fabrication (for Project 6), 206
    wiring diagram, 154
    wiring procedure (Project 3), 153
Circuit breakers, 65
Coaxial cable assembly procedure, 103
Coils. See Inductors
Conductors and cables, 17
Connectors
    audio, 61
    combination, 60
    flexible, flat, 59
    miniature, 59
    PC edge, 59, 148, 158, 160
    power line, 60
    rack and panel, 58, 147
    requirements, 58
    single-lead and multi-pin, 61
Contacts and contactors. See Switches
Crystals, 76

Desoldering
    procedure, 98
    process, 97
    tools, 13
Diagram
    block, 156
    chassis assembly, 144–151
    chassis wiring (Project 3), 154
    control panel layout (Project 4), 171
    schematic, 19
    sheet metal layout (Project 1), 131–133
    test set-up, 187, 225
Diodes (rectifiers)
    active elements, 39, 44
    package outline, 251
    power, 44
    small signal, 44
    vacuum tubes, 37
    zener, 45
Direct current (dc)
    sources, batteries and cells, 63

Edge connector, 164
Electrical shock, 2
Electronic devices
    semiconductor, 44
    vacuum tubes, 38
Electrostatic discharge (ESD)
    components susceptible to ESD (Table 1-3), 6
    defined, 3
    failure mechanisms, 5
    self-control program, 6
    static-safe work station, 7
    test circuit, 4
    typical ESD voltage levels (Table 1-1), 5
    typical prime charge sources (Table 1-2), 4

Field-effect transistors (FET)
    active elements, 46
    construction, 46
    IGFET, 46
    JFET, 46
    MOSFET, 46

Fire
- classification, 8
- extinguishing techniques, 8

First aid procedures, 2

Fuses, 64

Graphic symbols of common electronic components
- symbol reference guide, 229-241

Ground, 70

Hand tools, basic electronic
- adjustable wrench, 11
- diagonal cutters, 11
- needle nose pliers, 12
- nut drivers, 12
- safe use of, 9
- screwdriver, 12
- soldering aid, 12
- solder sucker, 13
- wire strippers, 13

IC regulator, 163

Inductors (coils), types of
- adjustable air-core and iron-core, 35
- combinations, 70, 72
- fixed air-core and iron-core, 34

Integrated circuits (ICs)
- standard packages, 263
- symbol outlines, 50

Lamps
- incandescent, 66
- neon, 66, 148

Latching relay circuit, 157, 161

Light-dependent-resistors (LDR). *See* Photodiodes

Logic circuits
- AND gates, 73
- delay lines, 75
- Exclusive OR, 75
- NOR gates, 74
- NOT and NAND gates, 74
- OR gates, 73
- oscillator, 75

Low voltage ac source, 162

Motors and generators, 68. *Also see* Rotating Machinery

Passive components
- capacitors, 29
- inductors (coils), 33
- mounting on PCB, 110
- resistors, 26
- transformers, 35

Photodiodes, 49

Plugs and Jacks, 58-63. *Also see* Connectors

Primary circuit, 159

Printed-circuit boards
- artwork development, 194
- assembly techniques, 108, 115, 179, 192, 200
- blanks, 18
- circuit design, 19, 175-182, 191-195, 209-221
- cleaning process, 113
- component lead preparation, 110
- fabrication, 195
- holders, 85
- rectifier board, 156
- repair, 115
- soldering process, 113
- tinning exercise, 88

Projects
- adjustable, bipolar regulated supply (Project 6), 202
- blinking LED circuit (Project 5), 188
- chassis assembly (Project 2), 140
- chassis fabrication (Project 1), 130
- chassis wiring (Project 3), 153
- continuity/voltage tester (Project 4), 168

Protection devices
- circuit breakers, 65
- electromagnetically actuated, 65
- fuses, 64, 151
- thermal breakers, 65

Phototype construction
- chassis assembly (Project 2), 140
- chassis fabrication (Project 1), 130
- chassis wiring (Project 3), 153

Reference designations, common consumer and military (Table 1-4), 21

Relays
- construction of, 55
- latching relay circuit, 157
- mounting socket, 148
- reed relay, 57
- solid-state relay, 57

Resistor color code system, 247

Resistors, types of
- fixed, 26
- metal film, 248
- special types, 28
- tapped and adjustable, 27
- variable, 27

Rotary switch, 54, 164

Rotating machinery, 68

Safety
- attitudes and environment, 2
- exam, 264-267
- general rules, 8

hand tools, 9
housekeeping, 9
machine tools, 10
personal protection, 10
soldering, 10
Schematic diagram
    adjustable, bipolar, regulated power supply (Project 6), 203
    blinking led circuit (Project 5), 189
    component symbols, 22, 118, 229–241
    continuity/voltage tester (Project 4), 169
    developing schematic diagrams, 122
    reading of, 158
    reference designations, 20
    rotary switch and edge connector (Project 3), 165
    symbol modifiers, 22
    test card (Project 3), 165
    test fixture (Project 3), 159
Semiconductor devices
    diodes, 44
    integrated circuits, 49
    special devices, 47
    transistors, 45
Sheet metal layout technique, 134, 206
Sheet metal fabrication tools
    90 degree notcher, 137
    drill press, 137
    hand drill motor, 138
    hand nibbler, 137
    hand punch, 136
    sheet metal brake, 138
    squaring shear, 134
    turret punch, 136
Sockets
    relay, 148
    vacuum tube, 39
Solder, type of, 15
Soldering
    basics of, 13
    defects of, 17
    desoldering, 97
    iron, 15
    printed-circuit boards, 113, 115
    process, 15
    solder feeding, 88
    tack soldering, 184
    tinning PCB, 88
    tinning stranded wire, 86
    wire splices, 87
    wire terminations, 85, 97
Solder wicking, 86
Speakers, headsets, earphones, and microphones. *See* Acoustical devices, 66–68
Special semiconductor devices
    IC regulator, 151, 157
    indicator outlines, 262
    light-emitting diodes, 49, 147
    photodiodes, 49
    SCR outlines, 259–260
    solar cells, 49
    thyristors (SCRs), 47
    triac, 48
    triac outlines, 261
    tunnel diodes, 48
    varactor diodes, 48
Specification sheet, development of
    for adjustable, bipolar-regulated power supply, 224
    for continuity/voltage tester, 186
Switches
    contact identification, 51
    DPDT, 53
    DPST, 52
    mechanical types, 51
    push-button, 54, 147
    SPDT, 52
    SPST, 52
    toggle, 51, 148
    wafer (rotary), 53, 146, 158, 164

Terminal blocks, 60, 149
Terminal boards, 62
    with terminal strips, 93
Terminal strips
    chassis mount, 151
    component mounting, 92, 159
    wire crimping, 92
Tester, continuity/voltage (Project 4), 168
Transformers, types of
    air-core, 36
    iron-core, 36
    powdered-iron-core, 36
    power (filament), 36, 149
Transistors
    bipolar-junction (BJT), 45
    field-effect (FET), 46
    package outlines, 252–258
    unijunction (UJT), 47
Twisted-shielded-pair cable, assembly procedure, 100

Unijunction transistor (UJT)
    active elements, 47
    construction, 47

Vacuum tubes
    cathode-ray (CRT), 42
    diodes, 37
    gas tubes, 41
    pentagrid, 40
    pentode, 39

photo tubes, 41
tetrode, 39
triodes, 39

Wire crimping
  flaws, 96
  procedure, 103
  soldering, 97
  terminal strips, 92
Wire splices
  rat-tail splice, 83
  soldering of, 87
  tee splice, 83
  Western-Union splice, 84
Wire strippers, types of
  adjustable, 81
  precision, 81
  thermal, 80
Wire stripping techniques
  shielded cable, 82
  solid wire, 81
  stranded wire, 81
Wires and cable, 69, 70
Wire wrapping, 242